I0494089

Disclaimer

The publisher of this book is by no way associated with the National Institute of Standards and Technology (NIST). The NIST did not publish this book. It was published by 50 page publications under the public domain license.

50 Page Publications.

Book Title: Annual Whole Building Energy Simulation of the NIST Net Zero Energy Residential Test Facility Design

Book Author: Joshua D. Kneifel;

Book Abstract: The National Institute of Standards and Technology (NIST) received funding through the American Recovery and Reinvestment Act (ARRA) to construct a Net Zero Energy Residential Test Facility (NZERTF). The initial goal of the NZERTF is to demonstrate
that a net-zero energy residential design can "look and feel" like a typical home in the Gaithersburg area. Demonstration phase of the project intends to demonstrate that the operation of the house does perform at "net zero," or produces as much electricity as it consumes over an entire year. The NZERTF is scheduled to begin the demonstration phase in 2013.

The purpose of this report is to create a whole building energy simulation that will replicate the NZERTF design to estimate its energy performance, both in aggregate as well as at the individual occupant and equipment level.

Citation: NIST TN - 1767

Keywords: Net zero energy construction; energy efficiency; residential building; whole building energy simulation

NIST Technical Note 1767

Annual Whole Building Energy Simulation of the NIST Net Zero Energy Residential Test Facility Design

Joshua Kneifel

http://dx.doi.org/10.6028/NIST.TN.1767

NIST

National Institute of Standards and Technology

U.S. Department of Commerce

NIST Technical Note 1767

Annual Whole Building Energy Simulation of the NIST Net Zero Energy Residential Test Facility Design

Joshua Kneifel
Applied Economics Office
Engineering Laboratory

http://dx.doi.org/10.6028/NIST.TN.1767

October 2012

U.S. Department of Commerce
Rebecca Blank, Acting Secretary

National Institute of Standards and Technology
Patrick D. Gallagher, Under Secretary of Commerce for Standards and Technology and Director

National Institute of Standards and Technology Technical Note 1767
Natl. Inst. Stand. Technol. Tech. Note 1767, 69 pages (October 2012)
http://dx.doi.org/10.6028/NIST.TN.1767
CODEN: NTNOEF

Abstract

The National Institute of Standards and Technology (NIST) received funding through the American Recovery and Reinvestment Act (ARRA) to construct a Net Zero Energy Residential Test Facility (NZERTF). The initial goal of the NZERTF is to demonstrate that a net-zero energy residential design can "look and feel" like a typical home in the Gaithersburg area. Demonstration phase of the project intends to demonstrate that the operation of the house does perform at "net zero," or produces as much electricity as it consumes over an entire year. The NZERTF is scheduled to begin the demonstration phase in 2013.

The purpose of this report is to create a whole building energy simulation that will replicate the NZERTF design to estimate its energy performance, both in aggregate as well as at the individual occupant and equipment level.

It is important to use a whole building energy simulation model to estimate the energy performance of the NZERTF for four main reasons. First, the building design needs to perform at net-zero to meet its demonstration phase goal. Second, the creation of a whole building energy simulation for the NZERTF can be used for sensitivity analysis. The building design and operation requires a plethora of assumptions on a number of levels, including the installed equipment, occupant behavior, and solar photovoltaic (PV) and solar thermal energy production. The ability to run "what-if" scenarios testing how each assumption impacts the energy performance of the NZERTF is beneficial for the residential construction industry. Third, the comparison of the energy simulation to the actual energy performance of the house will assist in fault detection during the demonstration phase of the NZERTF. Fourth, the comparison of the energy simulation to the actual energy performance of the house will assist the improvement of whole building energy simulation software.

Keywords

Net zero energy construction; energy efficiency; residential building; whole building energy simulation

Preface

This study was conducted by the Applied Economics Office (AEO) in the Engineering Laboratory (EL) at the National Institute of Standards and Technology (NIST). The study is designed to document the assumptions used to create the whole building energy simulation and the resulting estimated energy performance for the demonstration phase of the Net Zero Energy Residential Test Facility project. The intended audience includes researchers in the residential building sector concerned with net zero energy residential performance.

Disclaimer

Certain trade names and company products are mentioned in the text in order to adequately specify the technical procedures and equipment used. In no case does such identification imply recommendation or endorsement by the National Institute of Standards and Technology, nor does it imply that the products are necessarily the best available for the purpose.

The policy of the National Institute of Standards and Technology is to use metric units in all of its published materials. Because this report is intended for the U.S. construction industry that uses U.S. customary units, it is more practical and less confusing to include U.S. customary units as well as metric units. Measurement values in this report are therefore stated in metric units first, followed by the corresponding values in U.S. customary units within parentheses.

Acknowledgements

The author wishes to thank everyone involved in the NZERTF project. A special thanks to Farhad Omar, Dr. Hunter Fanney, Brian Dougherty, Dr. William Payne, Steve Emmerich, Dr. William Healy, Mark Davis, Piotr Domanski, Lisa Ng, Matthew Boyd, and Tania Ullah from NIST, Cathy Gates, Betsy Pettit, and Daniel Bergey from Building Science Corporation (BSC), and Richard Raustad and the other *E+* Help Desk team members for their assistance in designing the simulation model. Thank you to everyone for their advice and recommendations for the writing of this report, including Douglas Thomas and Dr. Robert Chapman of EL's Applied Economics Office, Mark Davis and Farhad Omar of EL's Energy and Environment Division, and Dr. Nicos S. Martys of EL's Materials and Structural Systems Division.

Author Information

Joshua D. Kneifel
Economist
National Institute of Standards and Technology
100 Bureau Drive, Mailstop 8603
Gaithersburg, MD 20899-8603
Tel.: 301-975-6857
Email: joshua.kneifel@nist.gov

Contents

List of Figures

List of Tables

List of Acronyms

Acronym	Definition
ACH	Air Changes Per Hour
AEO	Applied Economics Office
ARRA	American Recovery and Reinvestment Act
ASHRAE	American Society of Heating, Refrigerating and Air-Conditioning Engineers
BA	Building America
BSC	Building Science Corporation
BTP	Building Technology Program
CFA	Conditioned Floor Area
CFM	Cubic Feet Per Minute
COP	Coefficient of Performance
DHW	Domestic Hot Water
DOE	Department of Energy
E+	EnergyPlus
EERE	Energy Efficiency and Renewable Energy
EL	Engineering Laboratory
ETL	Electrical Testing Labs
FPW	Fraction of Peak Wattage
FPF	Fraction of Peak Flow
HVAC	Heating, Ventilating, and Air Conditioning
LBL	Lawrence Berkeley Laboratory
MAT	Mean Average Temperature
MEL	Miscellanous Electric Load
N_{BR}	number of bedrooms
NIST	National Institute of Standards and Technology
NREL	National Renewable Energy Laboratory
ODB	Outdoor Dry-Bulb
PV	Photovoltaic
RECS	Residential Energy Consumption Survey
SHGC	Solar Heat Gain Coefficient
TMY	Typical Meteorological Year
VT	Visible Transmittance

1 Introduction

1.1 Background and Purpose

The National Institute of Standards and Technology (NIST) received funding through the American Recovery and Reinvestment Act (ARRA) to construct a Net Zero Energy Residential Test Facility (NZERTF). The goals of the NZERTF are twofold. First, to demonstrate that a net-zero energy residential design can "look and feel" like a typical home in the Gaithersburg area. The demonstration phase of the project will demonstrate that the operation of the house does perform at "net zero." There are multiple definitions of a "net zero building," but for this report "net zero building" is defined as a building that produces as much electricity as it consumes over an entire year. Second, create a test facility in which a wide variety of building technology research can be performed. The NZERTF is scheduled to begin the demonstration phase in 2013.

It is important to estimate the energy performance of the NZERTF for a several reasons. First, the building design needs to perform at net-zero to meet its demonstration phase goal. Second, the creation of a whole building energy simulation for the NZERTF can be used for sensitivity analysis. The building design and operation requires a plethora of assumptions on a number of levels, including the installed equipment, occupant behavior, and solar photovoltaic (PV) and solar thermal energy production. The ability to run "what-if" scenarios that determine how each assumption impacts energy performance of the NZERTF is beneficial for the residential construction industry. Third, the comparison of the energy simulation to the actual energy performance of the house will assist in fault detection during the demonstration phase of the NZERTF. Fourth, the comparison of the energy simulation to the actual energy performance of the house will assist the improvement of whole building energy simulation software.

The purpose of this report is to create a whole building energy simulation that will replicate the NZERTF design and be able to estimate its energy performance, both at an aggregate level as well as at individual occupant and equipment levels.

1.2 Literature Review

The Department of Energy's (DOE) Energy Efficiency and Renewable Energy (EERE) Building Technologies Program (BTP) is responsible for funding research at the national laboratories for the Building America (BA) program. The Building America (BA) program has been at the forefront of research of low-energy single-family housing design through a variety of outlets, including the BA Best Practices Series, case studies for new construction and retrofits, and technical reports and fact sheets. Hendron and Engebrecht (2010) defines the BA house protocols to be implemented when simulating house energy performance.

1.3 Approach

This report documents the assumptions made to create a whole building energy simulation model in the *EnergyPlus* (*E+*) simulation software estimating the energy performance of the NZERTF.[1] The geometry, building envelope, and hard-wired lighting design as well as some energy performance requirements are based on the specifications defined by the NZERTF project's architectural firm, Building Science Corporation (BSC). Based on the BSC specifications, the contractor selected interior equipment and lighting to meet those specifications. Occupant behavior assumptions for the NZERTF are defined in a forthcoming document, Omar (Forthcoming). For some operating conditions, the model uses assumptions defined in Hendron and Engebrecht (2010). Additional documents that assist the model design are *ASHRAE 90.2-2007*, *ASHRAE 62.2-2010*, and the *ASHRAE Fundamentals Handbook*. Their use will be defined in detail where appropriate in the remainder of the document.

---

[1] Department of Energy (2010)

2 General Assumptions

The *EnergyPlus* (*E+*) whole building energy software was chosen to design the simulation for the NZERTF. Regardless of the building design, a number of general assumptions must be made for the *E+* simulation, including the location, run period, time steps, design day conditions, ground and underground temperatures, and water mains temperatures.

In order to run an annual simulation in *E+*, it is necessary to have a weather file defining a number of parameters throughout the year. A database of weather files based on a 15 year average of actual weather data, referred to as a Typical Meteorological Year (TMY), are available from the *E+* website. The NZERTF is located on the main NIST campus in Gaithersburg, MD (Latitude 39.07, Longitude -77.13, Elevation 106 m). The nearest location with an available TMY3 weather file, the third iteration of the TMY files, is the Washington-Dulles Airport (Latitude 38.95, Longitude -77.43, Elevation 82 m) and is based on data from years 1991 through 2005.

Running an *E+* simulation requires a run period and a time step. The run period represents which days of the year for which the simulation will run. An annual run period is selected because the goal of the simulation is to show how the NZETF performs over the year as a whole. The time step represents the frequency at which EnergyPlus runs the simulation. The time step used for the analysis is 1 minute. Since the length of some occupant activities, such as sink hot water use, are modeled in 1 minute increments, the 1 minute time step allows for precise calculations of total energy and hot water use throughout the year.

A simulation model can automatically size building energy systems to ensure the equipment will perform as needed across the variety of weather conditions that occur during an entire year. A location has winter and summer design days that represent the typical extreme conditions under which a building must effectively perform. The first factor defining a location's design days are the heating degree days and cooling degree days for that location. A location's heating (cooling) degree days for a single day are defined as the difference between the daily average actual outdoor dry-bulb temperature and the outdoor dry-bulb temperature required to maintain the desired heating (cooling) set point temperature inside the building (18.3 °C [65 °F] for both heating and cooling in this simulation). The total degree days for a location are summed over the entire year for each heating and cooling. For the Washington-Dulles Airport location, the heating degree days at base 18.3 °C (65 °F) is 4735 and the cooling degree days at base 18.3 °C (65 °F) is 1119. Most components in the NZERTF simulation are sized based on actual specifications of installed equipment, but there a few components that are currently autosized.

The second factor is how often the conditions in the building are allowed to not meet the desired conditions for both heating and cooling. In this simulation, it is assumed that for 350 hours (0.4 %) annually, the building does not meet each of its heating and cooling conditions. The design days used in the simulation are defined in *ASHRAE Fundamentals* (ASHRAE 2009) and described in Table 2-1.

Table 2-1 Washington-Dulles Design Day Conditions

Parameter	Winter	Summer
Max Dry-Bulb	-12.5 °C (9.5 °F)	34.0 °C (93.2 °F)
Daily Temperature Range	0.0 °C (0.0 °F)	11.6 °C (20.9 °F)
Humidity Indicating Conditions at Max Dry-Bulb	-12.5 °C (9.5 °F)	24.0 °C (75.2 °F)
Barometric Pressure	100,344 Pa	100,344 Pa
Wind Speed	2.7 m/s (8.9 ft/s)	4.1 m/s (13.5 ft/s)
Wind Direction	310 deg.	220 deg.
Sky Clearness	0 %	100 %
Rain Indicator	0	0
Snow Indicator	0	0
Month/Day	01/21	07/21
Humidity Indicating Type	Wet-Bulb	Wet-Bulb

The ground temperatures, both surface and underground, impact the heat transfer between the building and the ground and vary by the time of year. For the Washington-Dulles Airport location, monthly average surface ground temperatures vary from 19.71 °C (67.5 °F) to 22.49 °C (72.5 °F), as seen in Table 2-2.

Table 2-2 Washington-Dulles Average Monthly Ground Temperature

Month	Temperature (°C)	Temperature (°F)	Month	Temperature (°C)	Temperature (°F)
January	19.71	67.5	July	22.49	72.5
February	19.70	67.5	August	22.42	72.4
March	19.75	67.6	September	21.82	71.3
April	19.80	67.6	October	20.62	69.1
May	20.96	69.7	November	20.02	68.0
June	22.14	71.9	December	19.82	67.8

Underground temperatures for the exterior of the basement floor and basement walls must be calculated because each has a different average distance below the surface. The *E+ Basement Utility Preprocessor* calculates these time-varying temperatures based on the basement wall/floor construction and Washington-Dulles location, which is displayed in Table 2-3. The preprocessed values are called into the *E+* simulation file in a comma delimited format.

Table 2-3 Washington-Dulles Average Monthly Shallow Underground Temperature

| Month | Basement Temperature | | | |
| | Wall | | Foundation | |
	(°C)	(°F)	(°C)	(°F)
Jan.	19.40	66.9	21.88	71.4
Feb.	19.68	67.4	21.87	71.4
March	20.04	68.1	21.87	71.4
April	20.41	68.7	21.87	71.4
May	21.14	70.1	21.87	71.4
June	21.62	70.9	21.87	71.4
July	21.88	71.4	21.88	71.4
Aug.	21.89	71.4	21.88	71.4
Sept.	21.61	70.9	21.88	71.4
Oct.	21.14	70.1	21.89	71.4
Nov.	20.42	68.8	21.88	71.4
Dec.	19.87	67.8	21.88	71.4

The water mains temperature is the temperature of the water at the entrance point of the main water line into the NZERTF, which varies by the day of the year. $E+$ uses a correlation calculation method to estimate the temperature based on two parameters. The annual average outdoor air temperature is assumed to be 11.96 °C (53.5 °F). The maximum difference in outdoor air temperature is assumed to be 26.9 °C (80.4 °F). According to the $E+$ *Engineering Reference* documentation, "The correlation was developed by Craig Christensen and Jay Burch…(Hendron et al., 2004)," and is expressed in the following equation that is abstracted from the $E+$ documentation.

$$T_{mains} = \left(T_{out.avg} + 6\right) + ratio * \left(\Delta T_{out.maxdiff} / 2\right) * \sin(0.986 * (day - 15 - lag) - 90)$$

Where

T_{mains} = water mains temperature (°F)

$T_{out.avg}$ = average annual outdoor air temperature (°F)

$\Delta T_{out.maxdiff}$ = maximum difference in monthly average outdoor air temperature (°F)

Day = Julian day of the year (1-365)

Ratio = 0.4 + 0.01 * ($T_{amb.avg}$ – 44)

Lag = 35 – 1.0 * ($T_{amb.avg}$ – 44) (°F)

3 Building Design

The $E+$ model design of the NZERTF requires detailed definitions of its geometry, surface materials and constructions, shading surfaces, zoning, and internal mass. Each of these is defined in detail in this chapter.

3.1 Surfaces and Zones

The dimensions specified in the BSC architectural massing model, seen in Figure 3-1, are used along with Google SketchUp and NREL's Open Studio plug-in to construct the building geometry of the NZERTF. Total conditioned floor area of the $E+$ model is 284.6 m^2 (3063 ft^2) with 148.8 m^2 (1601.5 ft^2) on the first floor and 135.8 m^2 (1461.5 ft^2) on the second floor. Actual conditioned floor area of the NZERTF is 251.7 m^2 (2709 ft^2) with 141.0 m^2 (1518 ft^2) on the first floor and 110.6 m^2 (1191 ft^2) on the 2nd floor. There are two reasons the conditioned floor area of the simulation model is 32.9 m^2 (354 ft^2) greater than the actual house design. First, the $E+$ model does not account for the open foyer/stairway, which is approximately 19.0 m^2 (204 ft^2). Second, the gable walls (west wall and east wall) of the 2nd floor have built in storage under the gable, which decreases the conditioned floor area by approximately 14.3 m^2 (154 ft^2). These two aspects of model account for approximately 33.3 m^2 (358 ft^2), which decreases the conditioned floor area to 251.3 m^2 (2705 ft^2) or a difference of only 0.4 m^2 (4 ft^2). Even though these two aspects of the house are not considered conditioned floor area, they do add to the volume of space that will be conditioned.

Figure 3-1 BSC Architectural Massing Model

Figure 3-2 shows the Google SketchUp three-dimensional geometry of the $E+$ model for the NZERTF. The model includes seven separate zones with two conditioned (1st floor and 2nd floor) and five unconditioned zones (open web joist space between the 1st and 2nd floors, basement, main attic, living room attic, and patio). The front porch and detached garage with the covered walkway are all treated as shading surfaces, which block sunlight but do not impact the thermal performance of the building envelope. There are 172 building surfaces, 43 fenestration surfaces, and 409 shading surfaces in total.

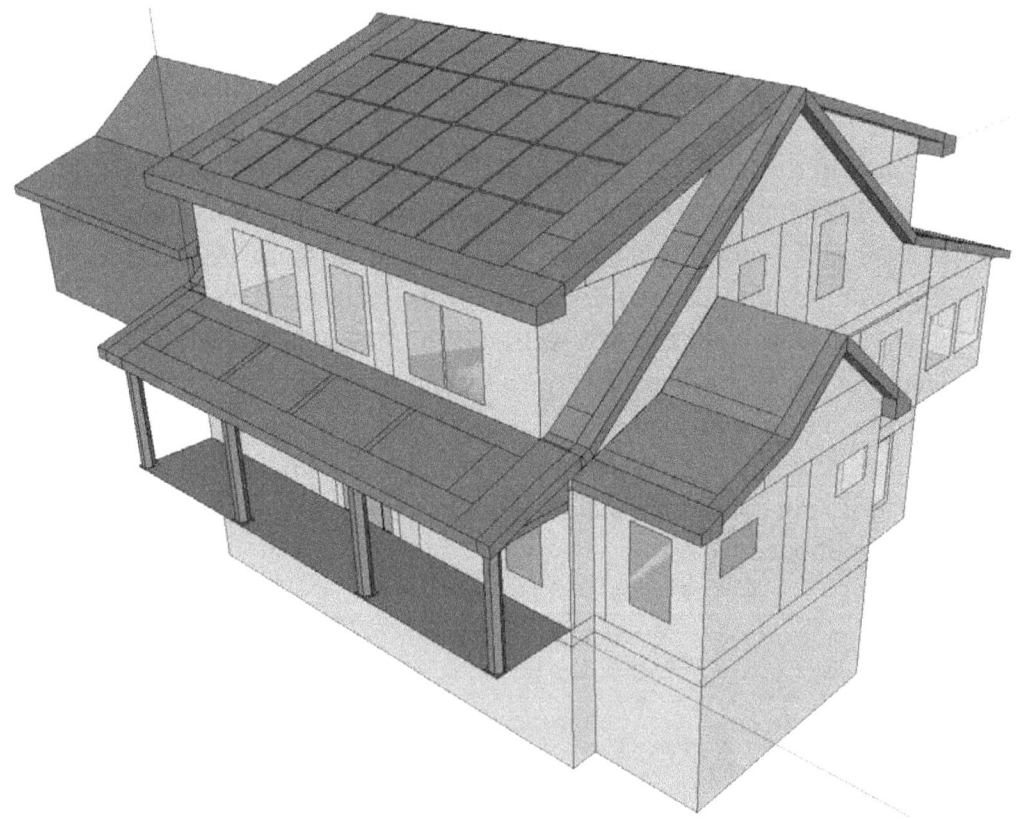

Figure 3-2 Google SketchUp 3-D Representation of the $E+$ Model

3.2   Building Materials and Constructions

A number of parameters must be defined for each $E+$ "surface" shown in Figure 3-2, including the zone in which the surface is located, what the outside of the surface is touching (another zone, the outdoors, or ground), and the surfaces "construction." A construction is a combination of one or more "materials."

Table 3-1 Material Parameter Examples

Field	Units	Obj1	Obj2	Obj3
Name		Asphalt shingles	½" Plywood Sheathing	5/8" Plywood Sheathing
Roughness		Very Rough	Smooth	Smooth
Thickness	M	0.0032	0.0127	0.0159
Conductivity	W/m-K	0.04	0.07	0.12
Density	Kg/m^3	1120	400	544
Specific Heat	J/kg-K	1260	1300	1210

Each material requires multiple parameters, including roughness, thickness, conductivity, density, and specific heat as shown in the example in Table 3-1. The characteristics of each material are collected from a variety of sources. The primary source of information is the *ASHRAE 2009 Handbook of Fundamentals* materials data. If the information was not available from this source, an internet search was used to find comparable materials and their characteristics.[2] Examples of "constructions" can be seen in Table 3-2. Constructions can have as few as one layer (a.k.a. "material") and up to more than ten layers.

Table 3-2 Construction Layers Examples

Field	Obj1	Obj2	Obj3
Name	First Floor Ceiling – Living Room	Roof Assembly – Stud	Roof Assembly – Insulation
Outside Layer	11 7/8" LSL Engineered Joist	Asphalt shingles	Asphalt shingles
Layer 2	½" GWB	Peel and Stick Roof Membrane	Peel and Stick Roof Membrane
Layer 3		5/8" Plywood Sheathing	5/8" Plywood Sheathing
Layer 4		Foil-faced Polyiso- 1.5"	Foil-faced Polyiso- 1.5"
Layer 5		Foil-faced Polyiso- 2.0"	Foil-faced Polyiso- 2.0"
Layer 6		Foil-faced Polyiso- 1.5"	Foil-faced Polyiso- 1.5"
Layer 7		Air Barrier Membrane	Air Barrier Membrane
Layer 8		½" Plywood Sheathing	½" Plywood Sheathing
Layer 9		**1 ¾" x 11 7/8" LSL Engineered Roof Rafters – Wood Only**	**1 ¾" x 11 7/8" LSL Engineered Roof Rafters – Cellulose Insulation Only**
Layer 10		½" GWB	½" GWB

Due to the thermal bridging caused by the wood framing of the facility, two construction types are necessary for each building surface (excluding fenestrations). An example can be seen for the two "constructions" used for the roof assembly in Table 3-2. Instead of precisely specifying the location of each piece of wood framing, the area of framing for each surface is totaled (15 %), and that portion of the surface is given the framing "construction" with the remainder of the

---

[2] Source: http://www.engineeringtoolbox.com/

surface given the non-framing "construction" (85 %).[3] The *ASHRAE Fundamentals Handbook* suggests using this approach to approximate the thermal bridging of wood framing while minimizing the time required for the calculations.[4]

The fenestration surface construction materials for windows are defined based on 3 simple parameters: U-factor, Solar Heat Gain Coefficient (SHGC), and Visible Transmittance (VT). This approach allows the rated window performance to be modeled while simplifying window "materials" and "constructions." The window parameters can be seen in Table 3-3, and are based on the minimum requirements specified in the BSC specifications.[5]

Table 3-3 Window Parameters Example

Field	Units	Values
Name		Baseline Windows
U-Factor	W/m^2-K	1.1356
Solar Heat Gain Coefficient		0.25
Visible Transmittance		0.40

Interior shades for the windows are based on Hunter Douglas Honeycomb Archiletta Daisy White shades. Shades have been included in the simulation, but they are assumed to never be used because they will not be used during the demonstration phase of the NZERTF project. Most of the parameter values, seen in Table 3-4, could not be obtained from the product specifications, which required the use of $E+$ example shade constructions for the last ten window shade parameters.

---

[3] This portion is assumed to be 15 % for the "advanced framing" approach. In reality this value could vary above or below 15 % (BSC, 2009).

[4] There are alternative approaches to estimate a surface's thermal performance. Instead of defining each material, the performance of the entire construction could be estimated or the average performance of all constructions for a particular group of surfaces.

[5] These parameters assume no difference in performance of the windows regardless of the window type (awning or double hung).

Table 3-4 Window Shade Parameters

Field	Values
Name	Honeycomb - White
Solar Transmittance	0.2
Solar Reflectance	0.7
Visible Transmittance	0.2
Visible Reflectance	0.7
Thermal Hemispherical Emissivity	0.9
Thermal Transmittance	0.0
Thickness	0.005
Conductivity	0.1
Shade to Glass Distance	0.05
Top Opening Multiplier	0.5
Bottom Opening Multiplier	0.5
Left-Side Opening Multiplier	0.5
Right-Side Opening Multiplier	0.5
Airflow Permeability	0.05

3.3   Shading Surfaces

The shading surfaces serve two purposes: (1) create shading for windows and walls and (2) create surfaces on which solar photovoltaic and solar thermal systems can be installed. All shading surfaces are assumed to block 100 % of light, and have no thermal impacts on the house excluding the blocking of natural light.

3.4   Internal Mass

ASHRAE 90.2-2007 sets the occupancy thermal mass (furniture and contents) at 39.1 kg/m^2 (8 lb/ft^2) of conditioned floor area based on 5.1 cm (2 in) wood with a specific heat of 1633 J/kg*°C (0.39 Btu/lb*1°F) and a conductivity of 0.144 W/m*K (1.0 Btu*in/h*ft^2*°F). Structural mass is set at 24.41 kg/m^2 (5 lb/ft^2) based on 1.27 cm (0.5 inch) gypsum board. Given these parameters, the internal thermal mass assumption is as follows for the two conditioned zones.

(1) 1st Floor
   a.   5811 kg (12 812 lb) of 5.1 cm (2 in) wood
   b.   3632 kg (8008 lb) of 1.27 cm (0.5 in) gypsum board
(2) 2nd Floor
   a.   5303 kg (11 692 lb) of 5.1 cm (2 in) wood
   b.   3315 kg (7308 lb) of 1.27 cm (0.5 in) gypsum board

E+ requires the definition of a "construction" and total area for each internal mass object. The constructions are assumed to be 5.1 cm (2 in) wood and 1.27 cm (0.5 in) gypsum board. It is assumed that the wood density is equivalent to white oak, or 16.02 kg/m^3 (47 lb/ft^3). So given the

5.1 cm (2 in) thickness of the wood, the kg/m^2 (lb/ft^2) is 38.2 (47/6=7.83). A 1.27 cm (0.5 in) gypsum board has a density of 10.16 kg/m^2 (2.08 lb/ft^2). Total internal masses for both internal mass materials on each floor are calculated below.

(1) 1st Floor
   a. 5811/38.2 (12 812 / 7.83) = 152 m^2 (1636 ft^2) of wood
   b. 3632/10.16 (8007.5 / 2.08) = 358 m^2 (3850 ft^2) of gypsum board
(2) 2nd Floor
   a. 5303/38.2 (11692 / 7.83) = 139 m^2 (1493 ft^2) of wood
   b. 3315/10.16 (7307.5 / 2.08) = 326 m^2 (3513 ft^2) of gypsum board

4 Building Use

The occupant's use of the NZERTF is just as important as the building envelope design when it comes to meet its annual net zero energy goal. The building components (e.g. interior equipment and lighting systems), occupant preferences (e.g. thermostat setpoints), and occupant behavior (e.g. occupancy, hot water use, and activity levels) all impact a house's energy performance.

4.1 Occupancy

The occupancy is assumed to be a family of four, two parents and two children (14 years old and 8 years old). The assumed occupant activity levels and the resulting sensible and latent heat gains shown in Table 4-1 are based on Hendron and Engebrecht (2010). The loads are assumed to be constant, which should be representative of the occupancy impacts, on average. There will be some variation depending on the actual activity of the occupants.

Table 4-1 Occupant Activity Level

Occupant Internal Load	kJ (Btu) Per Hour	
	1st Floor	2nd Floor
Sensible	243 (230)	221 (210)
Latent	200 (190)	148 (140)

Occupancy schedules for each of the four family members are based on a meticulously detailed 7 day narrative defined in Omar (Forthcoming). Figure 4-1 condenses the occupancy schedules to create an occupancy density by hour of each day of the week.

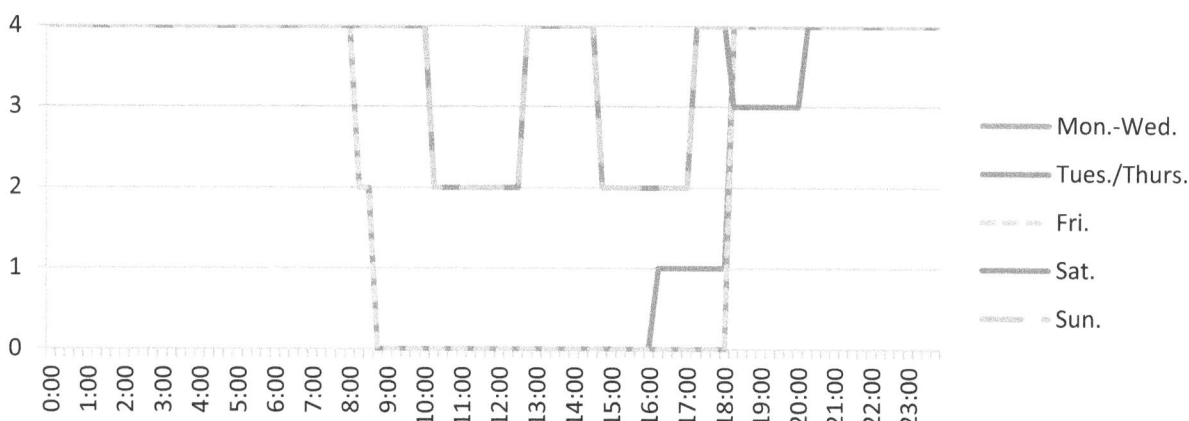

Figure 4-1 Occupancy Density

13

Figure 4-2 displays the occupancy schedules by hour for each day of the week for each occupant. All four occupants have their own unique weekly routine that controls for work, school, and extracurricular activities outside the home as well as various activities while occupying the NZERTF.

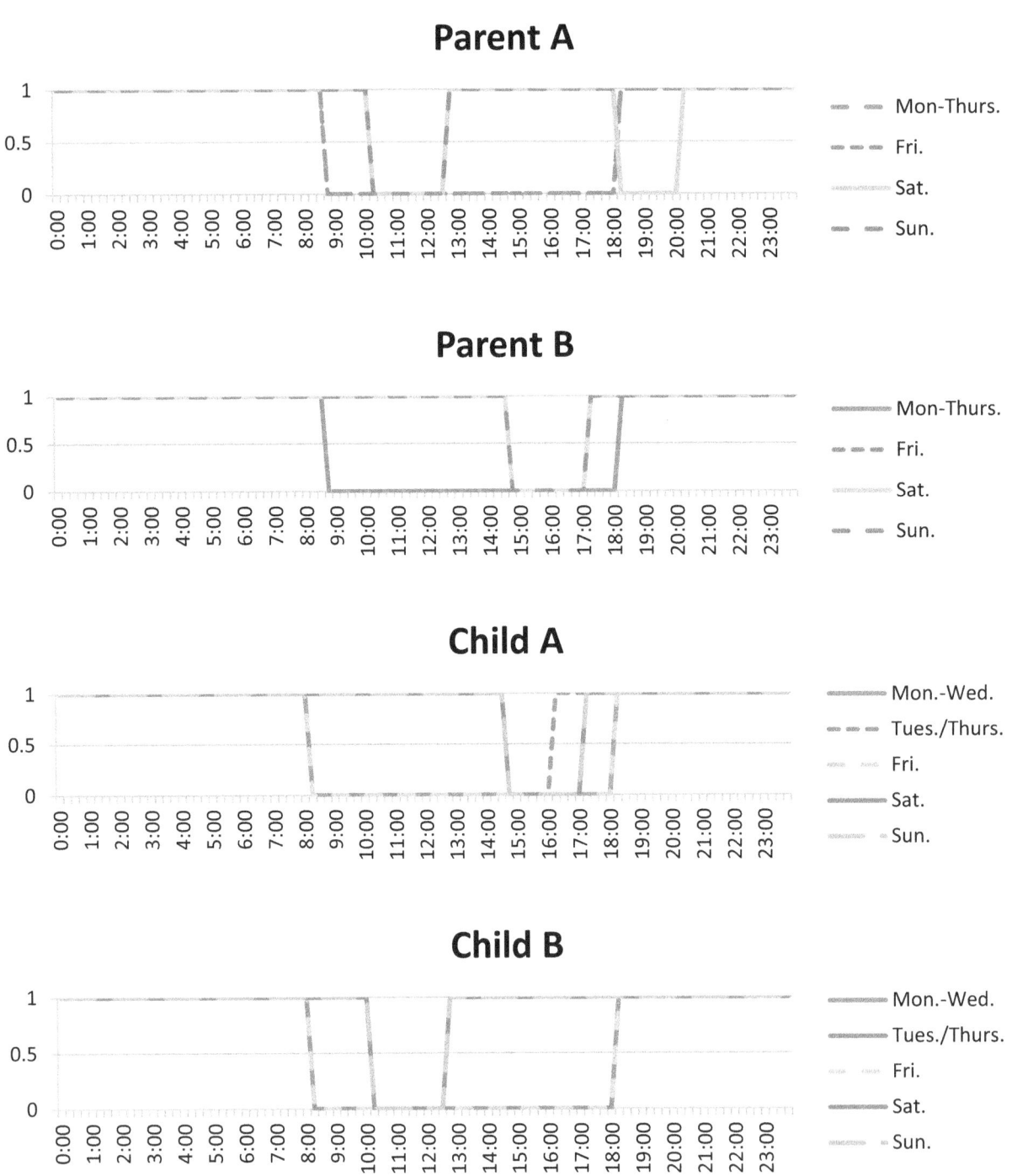

Figure 4-2 Occupancy Schedules by Occupant

4.2 Lighting

Electricity use and internal loads from interior lighting in the NZERTF are estimated based on Omar (Forthcoming) to determine occupancy by room, and then turning on all lights in the room while it is occupied. The sum of lighting wattage by room is shown in Table 4-2.

Table 4-2 Lighting Total Wattage by Room

Floor	Room	Watts
1st	Kitchen	107
	Dining Room	13
	Living Room	92
	Office	28
	1st Floor Bath	46
2nd	Master Bedroom	13
	2nd Bedroom	28
	3rd Bedroom	28
	Master Bathroom	72
	2nd Bathroom	24
Basement		217

The lighting schedules in terms of fraction of peak wattage (FPW) shown in Figure 4-3 are based on Omar (Forthcoming). Only some of the rooms in the conditioned space are assumed to be occupied during the narrative. For example, lights in the office and hallways are never turned on. Based on the narrative, the use of these areas should be minimal (a.k.a. a few seconds at a time).

1st Floor Bath

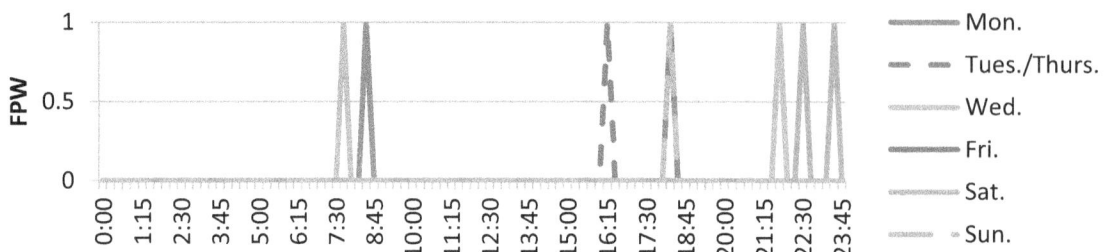

2nd Floor Bath

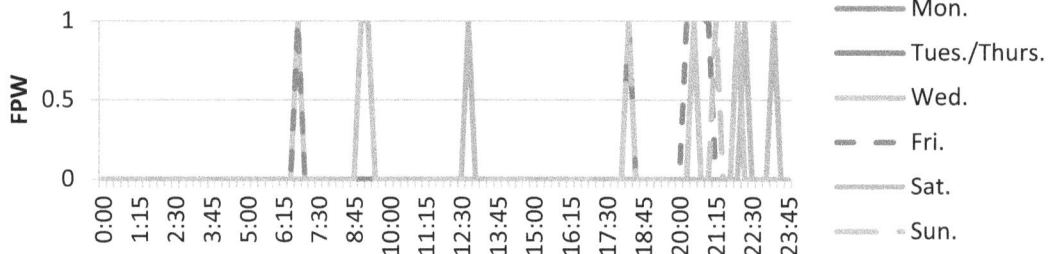

2nd Bedroom

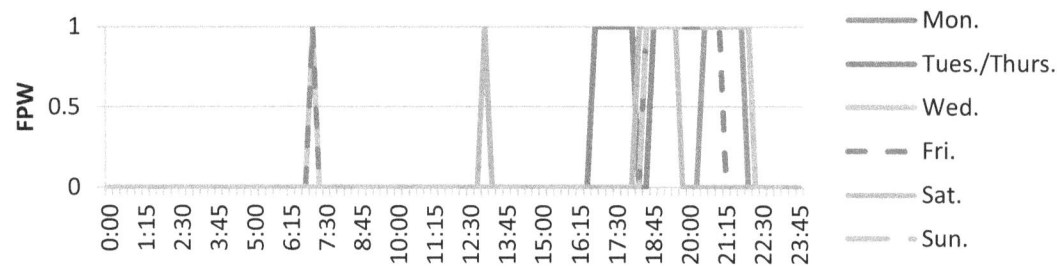

3rd Bedroom

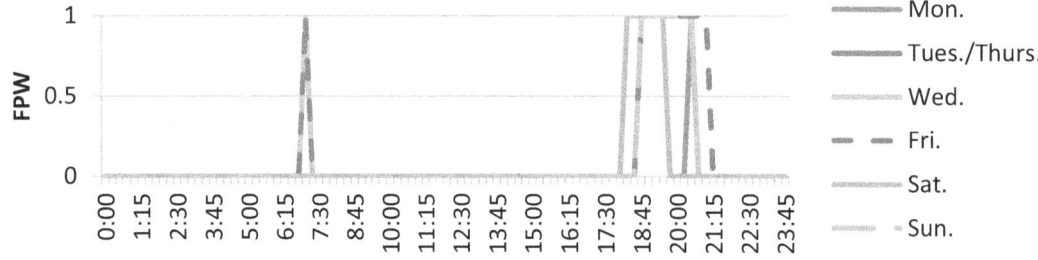

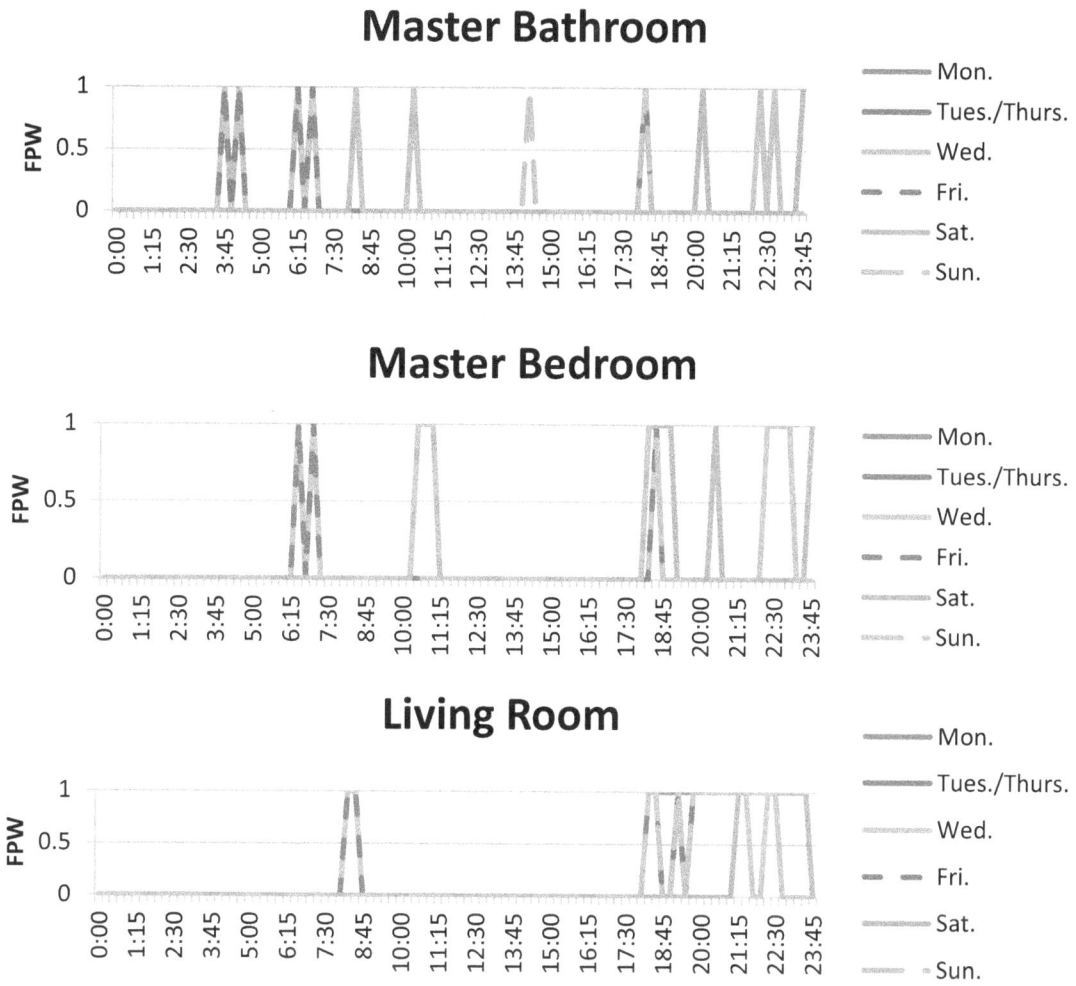

Figure 4-3 Lighting Schedules – Fraction of Peak Wattage

All other lighting is assumed to be zero. The lighting in the basement is currently assumed to never be used because it is neither finished nor occupied. Exterior lighting for the patio, garage, and the outdoor lights has been excluded. Garage and exterior lighting are not of a major concern because the lighting does not impact the thermal load of the NZERTF, and will only slightly increase electricity use once included.

4.3 Non-HVAC Interior Equipment

Non-HVAC interior equipment includes large appliances and any miscellaneous electrical loads (MELs), such as televisions, computers, hair dryers, etc. Table 4-3 shows the large appliances to be installed in the NZERTF by the contractor, their wattage, and the fraction of electricity used by the appliances that is converted into sensible and latent loads.[6] The Energy Star ratings are used to calculate the average wattage for the operation of the refrigerator, clothes washer, and

---

[6] The NZERTF will simulate the cooktop in a different manner. Once the approach is finalized, the model will be updated.

dishwasher. The dishwasher is rated at 234 kWh/year for 215 loads. Assuming a 1-hour cleaning cycle, the average wattage is 1090 W. The clothes washer is rated at 155 kWh/yr for 416 loads. Assuming a 45-minute cleaning cycle, the average wattage is 500 W. The refrigerator combines the Energy Star rated energy use (335 kWh) and the load profile from Hendron and Engebrecht (2010) to reverse engineer the peak wattage (45.7 W) to generate the target electricity use. The wattages of the clothes dryer and cooking equipment are based on the manufacturing specifications.[7]

Table 4-3 Appliance Wattage and Sensible and Latent Load Fractions

Appliance	Brand	Model	Average	Sensible Load	Latent Load
Refrigerator	Frigidaire	FPUI1888L	45.7	1.00	0.00
Clothes Washer	Whirlpool	WFW97HEX	500	0.80	0.00
Clothes Dryer	Whirlpool	WED97HEX	5200	0.15	0.05
Dishwasher	Bosch	SHX68E15UC	1090	0.60	0.15
Range – Oven	Wolf	SO30-2F/S-TH	5100	0.40	0.30
Range – Cooktop	Wolf	CT301/S	3600*	0.40	0.30
Range – Hood	Wolf	CTWH30	330	0.00	0.00
Microwave	Wolf	MWD30-2F/S	950	1.00	0.00
*Assumes the use of only 2 burners.					

The MELs listed in Table 4-4 are defined in Omar (Forthcoming) and include any item that is used in an "average" household is included in the $E+$ model.[8] The total annual electricity use for each MEL is used to reverse engineer the wattage for the equipment. The MELs can be grouped into constant loads and variable loads. The sensible and latent load fractions are based on Hendron and Engebrecht (2010). The sensible load is assumed to be split 50/50 with convection/radiant fraction.

Table 4-4 Miscellaneous Electrical Load Wattage and Sensible and Latent Loads

Location	Miscellaneous	Constant or	Watts	Sensible Load	Latent Load
Bathroom	Curling Iron	Variable	85	0.734	0.16
	Hair Dryer	Variable	1875	0.734	0.16
Kitchen	Blender	Variable	450	0.734	0.16
	Can Opener	Variable	70	0.734	0.16
	Coffee Maker	Variable	550	0.734	0.16
	Hand Mixer	Variable	250	0.734	0.16
	Toaster	Variable	1400	0.734	0.16
	Toaster Oven	Variable	1200	0.734	0.16
	Slow Cooker	Variable	25.64	0.734	0.16

---

[7] The clothes dryer is assumed to run at peak wattage the entire drying cycle, which likely overestimates electricity use. The range hood wattage is based initial equipment specifications, not the Wolf range hood.
[8] A particular MEL is included if the average number per household is greater than 0.5.

Location	Miscellaneous	Constant or	Watts	Sensible Load	Latent Load
Living Room	Television	Variable	62.2	0.734	0.16
	Blu-Ray	Variable	17	0.734	0.16
	Cablebox	Constant	17.48	0.734	0.16
	Clock	Constant	2.98	0.734	0.16
	Stereo	Constant	17.51	0.734	0.16
	Video Game System	Variable	26.98	0.734	0.16
Office	Desktop Computer	Variable	74	0.734	0.16
	Desktop Monitor	Variable	27.6	0.734	0.16
	Answering Machine	Constant	6.49	0.734	0.16
	Modem	Constant	2.01	0.734	0.16
	Inkjet Printer	Constant	4.46	0.734	0.16
	Wireless Router	Constant	24	0.734	0.16
	Vacuum	Variable	542	0.734	0.16
Master Bedroom	Heating Pad	Variable	32.97	0.734	0.16
	Television	Variable	45.36	0.734	0.16
	Blu-Ray	Variable	17	0.734	0.16
	Clock Radio	Constant	1.71	0.734	0.16
	Portable Fan	Variable	19.76	0.734	0.16
	2 Cell Phones	Constant	17.72	0.734	0.16
	Other	Constant	1.07	0.734	0.16
	Cablebox	Constant	17.48	0.734	0.16
2nd Bedroom	Boombox	Constant	1.92	0.734	0.16
	1 Cell Phone	Constant	8.86	0.734	0.16
	Clock Radio	Constant	1.71	0.734	0.16
	Laptop A	Variable	36.88	0.734	0.16
3rd Bedroom	Laptop B	Variable	36.8	0.734	0.16
Note: Sensible and latent load fractions are based on Hendron and Engebrecht (2010).					

Figure 4-4 shows the MEL use schedules by room defined in Omar (Forthcoming). Some rooms see fairly consistent occupancy behavior (kitchen) while others vary significantly throughout the week (living room).

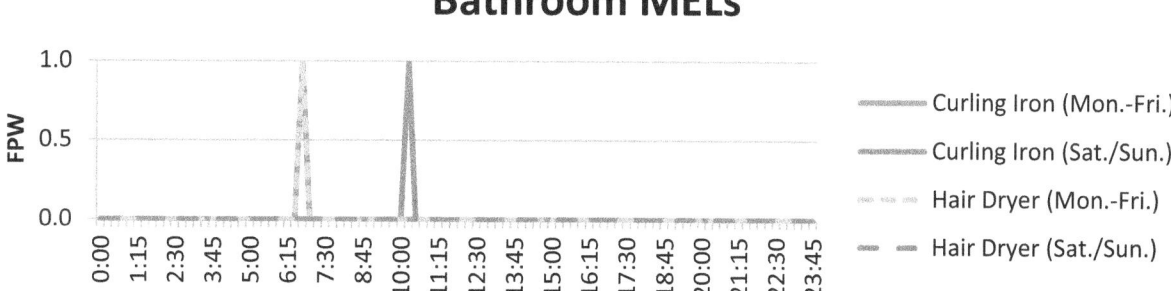

Master Bedroom MELs

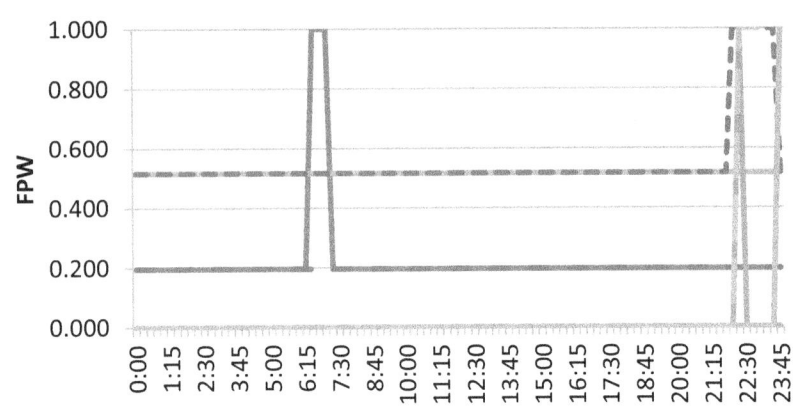

Legend:
- Master Bedroom TV (Mon.-Fri.)
- Master Bedroom TV (Sat./Sun.)
- Blu-Ray Master Bed (Mon.-Tues./Thurs.-Sat.)
- Blu-Ray Master Bed (Wed./Sun.)
- Heating Pad (Mon.-Thurs./Sun.)
- Heating Pad (Fri./Sat.)
- Portable Fan (Mon.-Thurs./Sun.)
- Portable Fan (Fri./Sat.)

Other Bedroom MELs

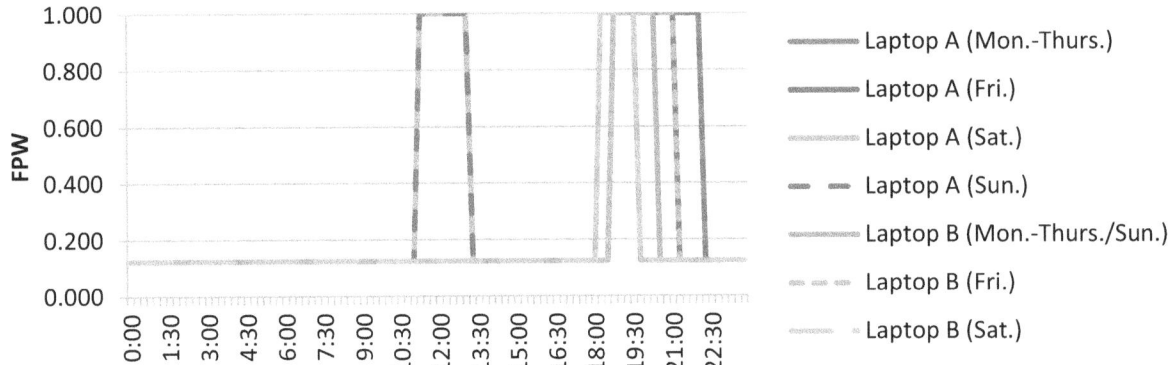

Legend:
- Laptop A (Mon.-Thurs.)
- Laptop A (Fri.)
- Laptop A (Sat.)
- Laptop A (Sun.)
- Laptop B (Mon.-Thurs./Sun.)
- Laptop B (Fri.)
- Laptop B (Sat.)

Living Room MELs

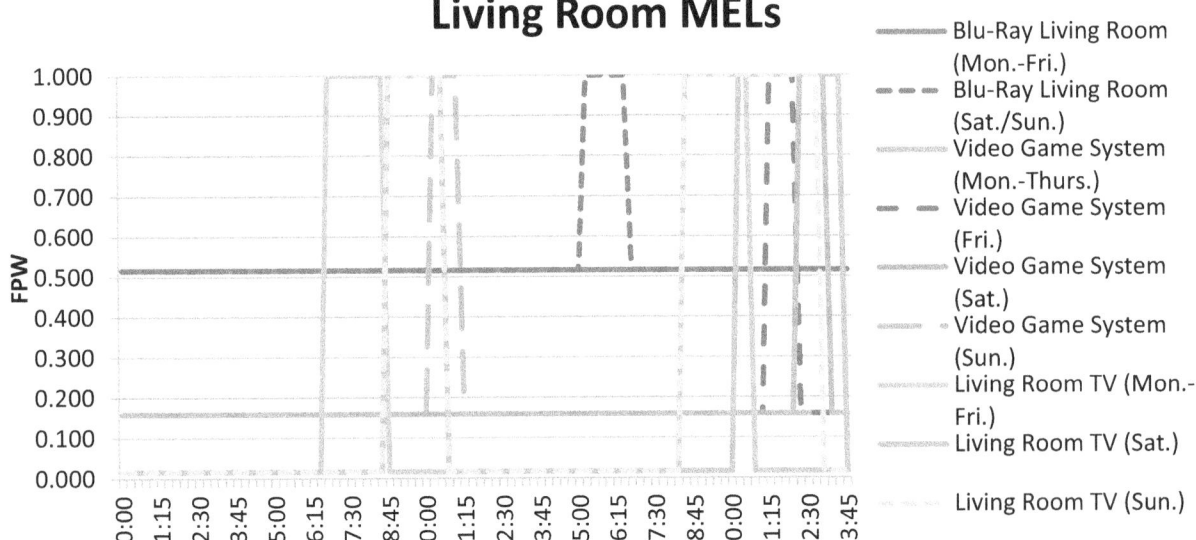

Legend:
- Blu-Ray Living Room (Mon.-Fri.)
- Blu-Ray Living Room (Sat./Sun.)
- Video Game System (Mon.-Thurs.)
- Video Game System (Fri.)
- Video Game System (Sat.)
- Video Game System (Sun.)
- Living Room TV (Mon.-Fri.)
- Living Room TV (Sat.)
- Living Room TV (Sun.)

Office MELs

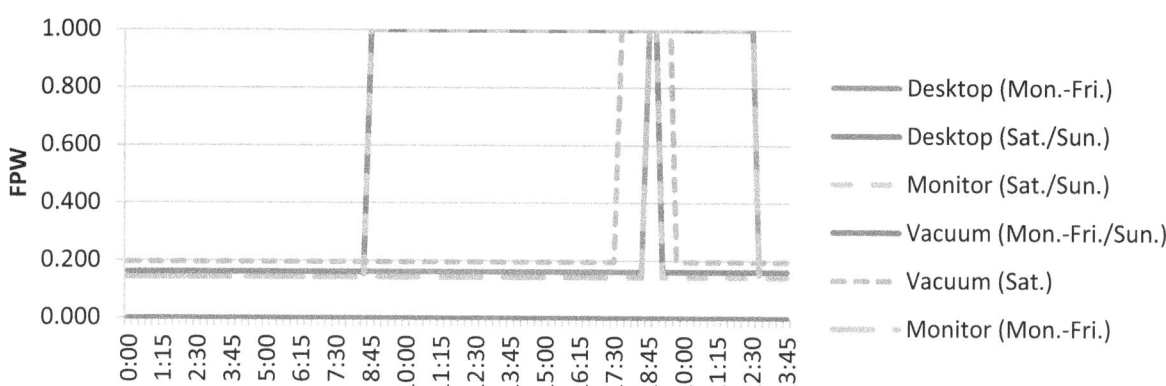

Legend:
- Desktop (Mon.-Fri.)
- Desktop (Sat./Sun.)
- Monitor (Sat./Sun.)
- Vacuum (Mon.-Fri./Sun.)
- Vacuum (Sat.)
- Monitor (Mon.-Fri.)

Kitchen MELs

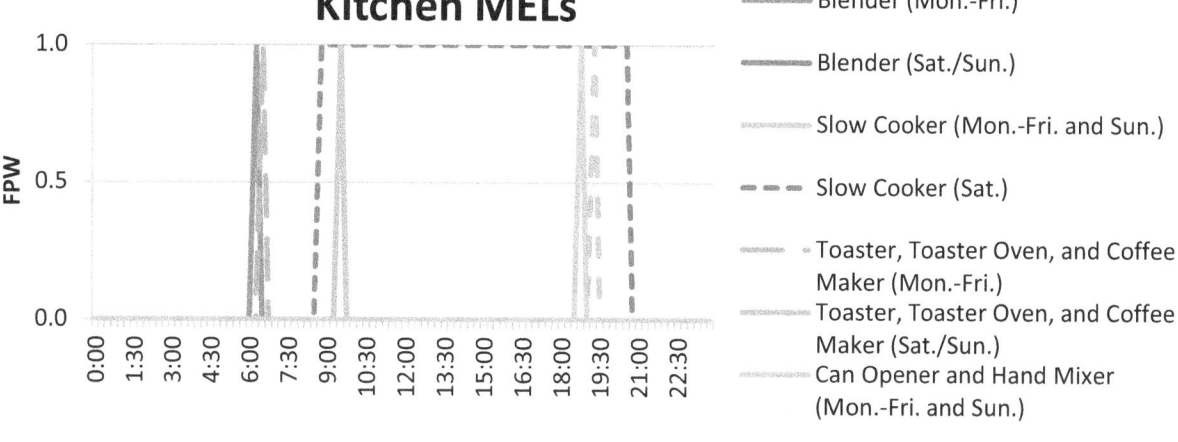

Legend:
- Blender (Mon.-Fri.)
- Blender (Sat./Sun.)
- Slow Cooker (Mon.-Fri. and Sun.)
- Slow Cooker (Sat.)
- Toaster, Toaster Oven, and Coffee Maker (Mon.-Fri.)
- Toaster, Toaster Oven, and Coffee Maker (Sat./Sun.)
- Can Opener and Hand Mixer (Mon.-Fri. and Sun.)

Kitchen Appliances

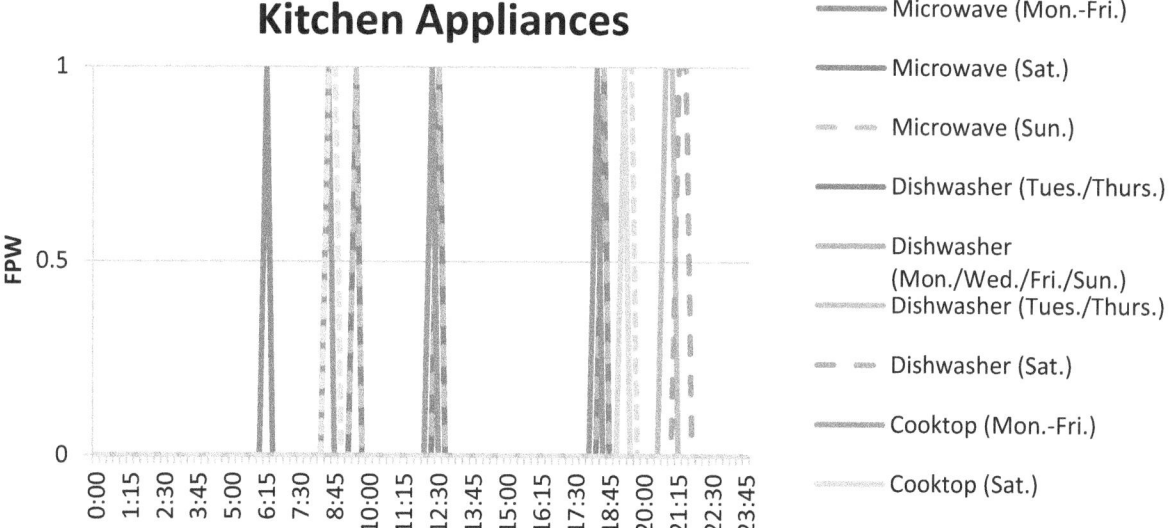

Legend:
- Microwave (Mon.-Fri.)
- Microwave (Sat.)
- Microwave (Sun.)
- Dishwasher (Tues./Thurs.)
- Dishwasher (Mon./Wed./Fri./Sun.)
- Dishwasher (Tues./Thurs.)
- Dishwasher (Sat.)
- Cooktop (Mon.-Fri.)
- Cooktop (Sat.)

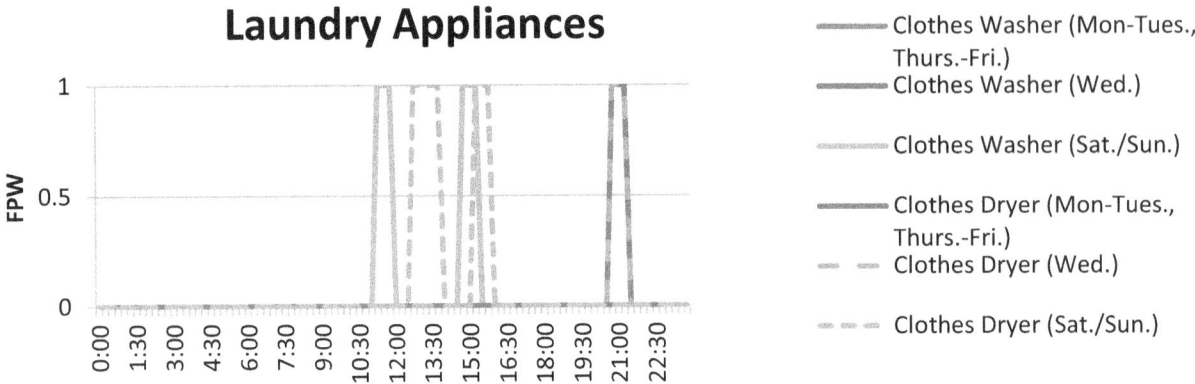

Figure 4-4 MEL Schedules – Fraction of Peak Wattage

4.4 Heating, Ventilation, and Air Conditioning

The *E+* model must specify all aspects of an HVAC system and the conditions to which the system must perform, including the thermostat setpoints, infiltration and ventilation rates, humidity controls, and HVAC equipment specifications. Each of these is defined in this section.

4.4.1 Thermostat

The thermostat setpoints have not yet been chosen for the demonstration phase of the NZERTF project. The thermostat settings selected for the *E+* model are shown in Table 4-5. These setpoints are based on protocols defined in Hendron and Engebrecht (2010) for the NZERTF's location. The setpoints for heating and cooling vary by the occupancy, time of the day, and day of the year. Heating setpoints are assumed to be 22.3 °C (72.1 °F) while occupied during the day, 20.1 °C (68.1 °F) while occupied during the nighttime, and 18.4 °C (65.1 °F) while unoccupied during the day. Cooling setpoints are assumed to be 23.6 °C (74.4 °F) while occupied and 26.3 °C (79.4 °F) while unoccupied. Nighttime occupancy is assumed to be between 12:00 midnight and when the first occupant wakes up (6:00 AM or 8:15 AM depending on the day of the week).

Table 4-5 Thermostat Setpoints

HVAC Condition	Setpoint C (F)		
	Occupied-Day	Unoccupied- Day	Occupied-Night
Heating	22.3 (72.1)	18.4 (65.1)	20.1 (68.1)
Cooling	23.6 (74.4)	26.3 (79.4)	23.6 (74.4)

The NZERTF is considered occupied anytime the occupancy density is greater than 0. Figure 4-5 charts the heating and cooling setpoints during each day of the week.

Heating Setpoint (°C)

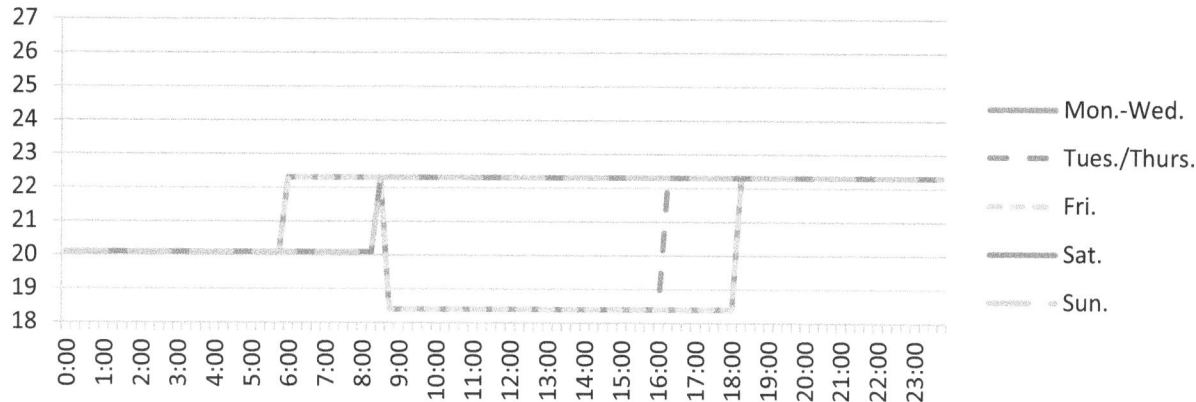

Cooling Setpoint (°C)

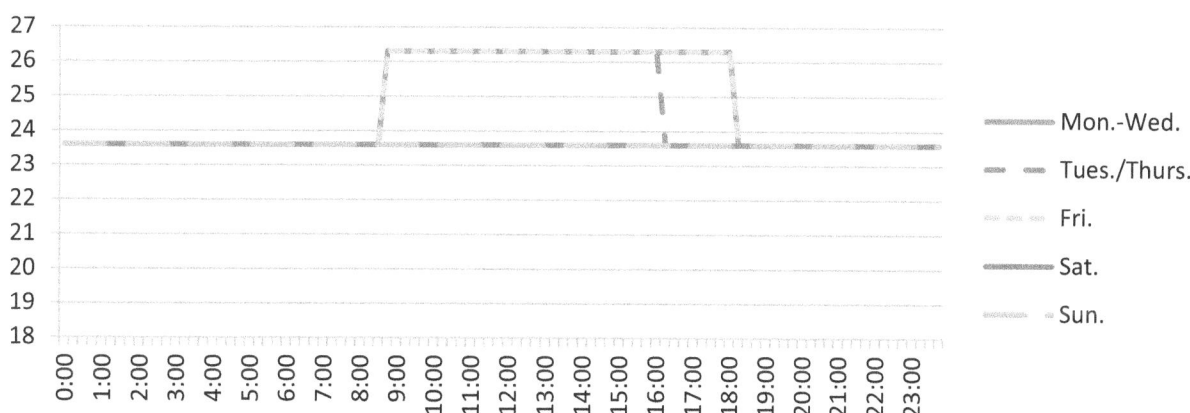

Figure 4-5 HVAC Heating and Cooling Setpoints

The heating and cooling seasons are based on the approach defined in Hendron and Engebrecht (2010). Heating is enabled for all months with monthly average temperatures (MATs) less than 19 °C (66 °F). Cooling is enabled for all months with an MAT equal to or greater than 19 °C (66 °F). An additional month is added to the beginning of the cooling season. Table 4-6 shows that the heating season is 8 months (October through May) while the cooling season is 5 months (May-September). The unique "tight" design of the NZERTF leads to higher than normal temperatures inside the house. As a result, it is necessary to expand the cooling season to include April and October to ensure the house remains in a comfortable temperature range. Also, the heating season is shortened to exclude May so that the heating and cooling seasons overlap by a single month.[9]

---

[9] The decision to expand cooling to April and October and remove May from the heating season was based on two factors. First, the simulation results using the initial approach led to a significant amount of time in April and

Table 4-6 Heating and Cooling Seasons

Season	Monthly Average Temperature C (F)											
	Jan.	Feb.	March	April	May	June	July	Aug.	Sept.	Oct.	Nov.	Dec.
Heating	-1 (31)	1 (34)	7 (44)	11 (52)	17 (62)	21 (70)	23 (75)	23 (74)	19 (66)	13 (55)	7 (45)	2 (35)
Cooling	-1 (31)	1 (34)	7 (44)	11 (52)	17 (62)	21 (70)	23 (75)	23 (74)	19 (66)	13 (55)	7 (45)	2 (35)
Source: http://countrystudies.us/ (2008)												

4.4.2 Infiltration and Ventilation

The NZERTF specifications from BSC include a target envelope tightness of 1 air change per hour based on a blower door test at 50 Pa of air pressure (ACH_{50}). The most recent air leakage test, performed by Everyday Green, resulted in a whole house air tightness of 0.215 m^3/s at 50 Pa, 456 cubic feet per minute at 50 Pa (CFM_{50}), or 0.61 ACH_{50}.[10] In order to account for this envelope airtightness in $E+$, it must be converted to either an infiltration rate in air changes per hour or an effective leakage area (ELA), in both cases at a specific pressure difference. ELA is the area of an orifice with a discharge coefficient of 1.0 that would allow the same amount of airflow through it as that measured through the entire building envelope during the pressurization test and is usually determined at 4 Pa.[11]

The approach chosen to model the building envelope in the occupied zones of the NZERTF is effective leakage area (ELA). The whole building leakage test estimates the ELA to be 189.0 cm^2 (29.3 in^2), and is split between the 1st floor and 2nd floor based on occupied floor volume. The 1st floor accounts for 52.3 % of the occupied volume while the 2nd floor accounts for the remaining 47.7 %, which leads to an ELA of 98.8 cm^2 (15.3 in^2) and 90.2 cm^2 (14.0 in^2), respectively.

All infiltration is assumed to occur in the occupied zones while the unoccupied zones in the conditioned space (basement, open web joists, and attic space) have no infiltration. The basement is fully underground and will only have infiltration through the egress window. The open web joists have minimal surface area shared with the exterior building envelope. The attic space may have some air leakage, but its leakage is grouped in with the 2nd floor. The patio is not in the conditioned space, and will not impact the heating and cooling energy use.

A blower door test is performed to determine the envelope airtightness. It does not account for the effect of opening of windows and doors (as a result of occupant activity) on infiltration. *ASHRAE 90.2-2007* assumes 0.15 ACH due to exhaust fans and occupants opening and closing

---

October for which the cooling setpoint temperatures were not met. Second, the tight building design will maintain higher temperatures during cool months, which makes it unnecessary to heat the NZERTF during the month of May.
[10] Everyday Green (2012).
[11] Source: ASRHAE Fundamentals (2009) – Chapter 16

of exterior doors and windows. Based on the way in which the NZERTF will be operated, there will be minimal occupant activity. For this reason, the model assumes no infiltration due to occupant activity. Nevertheless, the model was run with and without 0.15 ACH for occupant activity and it resulted in an increase of 1825 kWh (18 %) in energy use relative to the model with no occupant activity credit, which emphasizes the importance of correctly controlling for building occupancy. Table 4-7 shows the parameters used to simulate air infiltration in the $E+$ model. The stack coefficient controls for the hydrostatic pressure resulting from changes in air density while the wind coefficient controls for the static pressure exerted by wind on the building.[12]

Table 4-7 Infiltration Rates

Name	1st Floor	2nd Floor
ELA (cm²)	98.8	90.2
Stack Coeff.	0.00029	0.00029
Wind Coeff.	0.000325	0.000325

BSC used *ASHRAE Standard 62.2* to determine the minimum required mechanical outdoor air flow rate for the entire building to be 0.039 m³/s (83 CFM). $E+$ requires a mechanical ventilation rate for each zone. The 1st floor has 52.3 % of the volume while the 2nd floor has 47.7 % of the volume of the occupied space. Based on these values, the required minimum outdoor air flow rates for each zone can be calculated based on a weighted fraction of the whole house mechanical ventilation as 0.021 m³/s (20.5 L/s or 43.4 CFM) for the 1st floor, and 0.019 m³/s (18.7 L/s or 39.6 CFM) for the 2nd floor. These values will be updated with actual mechanical ventilation rates once operation of HRV is verified. Currently, exhaust fans (bathroom fans or range hoods) is not included in the model for simplicity.

4.4.3 HVAC Equipment

Heating and air conditioning are accomplished with a multispeed air-to-air heat pump with dehumidification-only mode. Mechanical ventilation is delivered through a dedicated heat recovery ventilator (HRV). There is also a whole-house dehumidifier. The heat pump and the HRV each have their own dedicated ductwork. The whole house dehumidifier will not be used during the demonstration phase because it is duplicative. $E+$ does not currently allow a multispeed air-to-air heat pump to run in dehumidification-only mode, requiring modeling of separate equipment to dehumidify the occupied space. However, it is not currently possible to simulate a whole house dehumidifier in $E+$. As a result, the $E+$ model simulates a dehumidifier in each of the occupied zones (1st floor and 2nd floor). $E+$ does not allow more than one whole house air loop, which forces the splitting of the HRV system into two zone-level HRV systems. Figure 4-6 shows the HVAC system modeled in the $E+$ simulation.

---

[12] For more details regarding the stack and wind coefficients, see the ASRHAE Fundamentals Handbook.

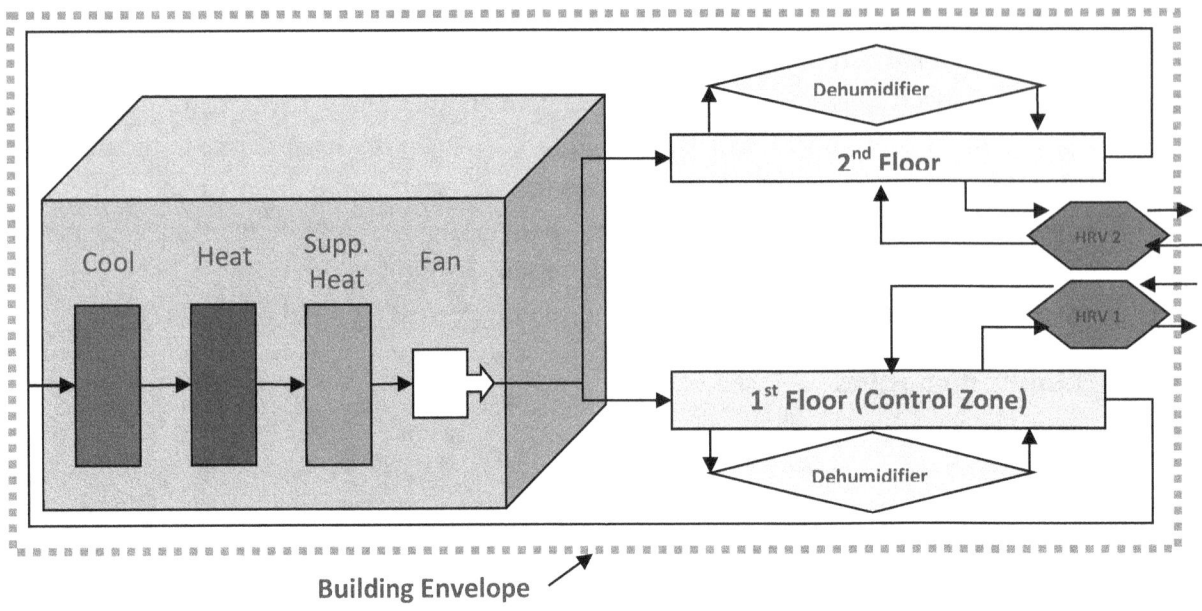

Figure 4-6 HVAC System Layout

The HVAC air-to-air heat pump is properly sized based on the assumed design day conditions (i.e. oversized by 0 %). There are only 2 zones that require their temperature to be controlled by the HVAC equipment, the 1st floor and 2nd floor. The 1st floor is used as the controlling zone, or the location of the thermostat. This will result in a floating of the 2nd floor air temperatures. The basement is conditioned, but since the space is not finished and will not be occupied during the demonstration phase, it is not necessary for the setpoint temperatures to be met in the basement. The open web joist space and attic space are in the conditioned space, but there is supply air from ductwork entering those spaces and the temperatures are allowed to fully float. The E+ input list in Table 4-8 are made for the sizing of thermal loads for each conditioned zone. Note that there is no outdoor air flow for the heating and cooling system. All mechanical ventilation of outdoor air is controlled by the HRVs.

Table 4-8 Zone Sizing Parameters

Parameter	Units	1st Floor	2nd Floor	Basement
Zone Cooling Design Supply Air Temperature	°C (°F)	12 (54)	12 (54)	12 (54)
Zone Heating Design Supply Air Temperature	°C (°F)	30 (86)	30 (86)	30 (86)
Zone Cooling Design Supply Air Humidity	kg-H2O/kg-air	0.008	0.008	0.008
Zone Heating Design Supply Air Humidity	kg-H2O/kg-air	0.008	0.008	0.008
Outdoor Air Method		Flow/Zone	Flow/Zone	Flow/Zone
Outdoor Air Flow Per Zone	m³/s	0.0	0.0	0.0
Cooling Design Air Flow Method		Design Day	Design Day	Design Day
Heating Design Air Flow Method		Design Day	Design Day	Design Day

The $E+$ input list in Table 4-9 are for the sizing of the air-to-air heat pump, which includes the fan and coils. The outdoor air flow rate is autosized as the sum of the zone-specific outdoor air flow rates, which totals to zero because the HRV system will meet all outdoor air requirements.

Table 4-9 HVAC System Sizing Parameters

Field	Units	Value
Type of Load to Size On		Sensible
Design Outdoor Air Flow Rate	m³/s	Autosize
Minimum System Air Flow Ratio		0.40
Preheat Design Temperature	°C (°F)	7 (45)
Preheat Design Humidity Ratio	kg-H2O/kg-air	0.008
Precool Design Temperature	°C (°F)	25 (77)
Precool Design Humidity Ratio	kg-H2O/kg-air	0.008
Central Cooling Design Supply Air Temperature	°C (°F)	12 (54)
Central Heating Design Supply Air Temperature	°C (°F)	30 (86)
Sizing Option		Non-Coincident
100 % Outdoor Air in Cooling		No
100 % Outdoor Air in Heating		No
Central Cooling Design Supply Air Humidity Ratio	kg-H₂O/kg-air	0.008
Central Heating Design Supply Air Humidity Ratio	kg-H₂O/kg-air	0.008
Cooling Design Air Flow Method		Design Day
Heating Design Air Flow Method		Design Day

The HVAC fan is a constant volume draw-through fan and has the E+ inputs listed in Table 4-10.

Table 4-10 HVAC Fan Parameters

Availability	Always Available
Fan Efficiency	70 %
Pressure Rise	125 Pa
Maximum Flow Rate	0.42 m³/s
Motor Efficiency	90 %
Motor in Airstream Fraction	1.0

The NZERTF will have a 2-ton AAON heat pump that has 2 speeds with gas reheat for dehumidification control. The following tables will define the parameters for a the 2-ton, 2-speed heat pump based on Electrical Testing Labs (ETL) test data obtained from a NIST HVAC expert. These values will be updated once the system is installed and tested.

Table 4-11 shows the $E+$ inputs for the cooling coil, which is a multispeed air-cooled electric DX coil. The cooling coil is assumed to have two speeds, referred to as "low speed" and "high speed." At low speed, the coil capacity is 5483 W with a rated coefficient of performance (COP) of 3.73 and a rated air flow rate of 0.23 m³/s (487 CFM). At high speed, the coil capacity is 7751 W with a rated COP of 3.69 and a rated air flow rate of 0.42 m³/s (890 CFM).

Table 4-11 Cooling Coil

Field	Units	Values
Availability		Always Available
Condenser Type		Air Cooled
Apply Part Load Fraction to Speeds Greater than 1		No
Apply Latent Degradation to Speeds Greater than 1		No
Fuel Type		Electricity
Number of Speeds		2
Speed 1 Rated Total Cooling Capacity		5483
Speed 1 Rated Sensible Heat Ratio		0.7
Speed 1 Rated COP		3.73
Speed 1 Rated Air Flow Rate		0.23
Speed 1 Rated Waste Heat Fraction of Power Input		0.1
Speed 1 Evaporative Condenser Effectiveness		0.9
Speed 2 Rated Total Cooling Capacity	W	7751
Speed 2 Rated Sensible Heat Ratio		0.7
Speed 2 Rated COP		3.69
Speed 2 Rated Air Flow Rate	m³/s	0.42
Speed 2 Rated Waste Heat Fraction of Power Input		0.1
Speed 2 Evaporative Condenser Effectiveness		0.9

The cooling coil performance curve types can be found in Table 4-12. I am not reporting the specifics of each curve here because of their complexity. Details on the functions are available upon request.

Table 4-12 Cooling Coil Performance Curves

Cooling Coil Performance Curve Type	Name	Form
Total Cooling Capacity Function of Temp. Curve	Heat Pump Cool Coil Cap-FT	Biquadratic
Total Cooling Capacity Function of Flow Fraction Curve	Heat Pump Cool Coil Cap-FF	Quadratic
Energy Input Ratio Function of Temp. Curve	Heat Pump Cool Coil EIR-FT	Biquadratic
Energy Input Ratio Function of Flow Fraction Curve	Heat Pump Cool Coil EIR-FF	Quadratic
Part Load Fraction Correlation Curve	Heat Pump Cool Coil PLF	Quadratic
Waste Heat Function of Temperature Curve	Waste Heat-FT	Biquadratic

The heating coil E+ inputs can be found in Table 4-13. The heating coil is a multispeed electric DX coil, and as with the cooling coil, the heating coil is assumed to have two speeds, referred to as "low speed" and "high speed." At low speed, the coil capacity is 4908 W with a rated COP of

28

4.02 and a rated air flow rate of 0.21 m³/s (487 CFM). At high speed, the coil capacity is 7675 W with a rated COP of 4.19 and a rated air flow rate of 0.42 m³/s (890 CFM).

Table 4-13 Heating Coil

Field	Units	Value
Availability		Always Available
Minimum ODB Temp. for Compressor Operation	°C	-17
Crankcase Heater Capacity	W	0
Maximum ODB Temp. for Crankcase Heater Operation	°C	10
Maximum ODB Temp. for Defrost Operation	°C	7.22
Defrost Strategy		Reverse Cycle
Defrost Control		On Demand
Defrost Time Period Fraction		0.058333
Resistive Defrost Heater Capacity	W	Autosize
Apply Part Load Fraction to Speeds Greater than 1		No
Fuel Type		Electricity
Number of Speeds		2
Speed 1 Rated Total Heating Capacity	W	4908
Speed 1 Rated COP		4.02
Speed 1 Rated Air Flow Rate	m³/s	0.21
Speed 1 Rated Waste Heat Fraction of Power Input		0.1
Speed 2 Rated Total Heating Capacity	W	7675
Speed 2 Rated COP		4.19
Speed 2 Rated Air Flow Rate	m³/s	0.42
Speed 2 Rated Waste Heat Fraction of Power Input		0.1

The heating coil performance curve types can be found in Table 4-14. The specifics of each curve are not reported here because of their complexity. Function details are available upon request.

Table 4-14 Heating Coil Performance Curves

Heating Coil Performance Curve Category	Name	Form
Total Heating Capacity Function of Temp. Curve	Heat Pump Heat Coil Cap-FT	Cubic
Total Heating Capacity Function of Flow Fraction Curve	Heat Pump Heat Coil Cap-FF	Cubic
Energy Input Ratio Function of Temp. Curve	Heat Pump Heat Coil EIR-FT	Cubic
Energy Input Ratio Function of Flow Fraction Curve	Heat Pump Heat Coil EIR-FF	Quadratic
Part Load Fraction Correlation Curve	Heat Pump Heat Coil PLF	Quadratic
Waste Heat Function of Temperature Curve	Waste Heat-FT	Biquadratic
Defrost Energy Input Ratio Function of Temp. Curve	Heat Pump Heat Coil DefCap-FT	Biquadratic

The supplemental heating coil is an electric resistance heating element with an efficiency of 1.0 and an autosized capacity. The operation of the NZERTF will attempt to minimize the need for the supplemental heating element.

The HRV system in the NZERTF is a VENMAR AVS HRV EKO 1.5 air-to-air flat plate heat exchanger. The HRV transfers heat between the exhaust air and supply air to decrease the heating and cooling load impact of the ventilation air. The HRV is the sole source of mechanical ventilation and operates year-round, 24 hours a day. The effectiveness of the HRV varies by the air flow rate, and is assumed to be the same for heating and cooling. Exhaust air recirculation is used to control for frost. The air flow rates through each HRV are based on the outdoor air requirement for each zone defined in Section 4.4.2. The E+ input values for the HRVs are listed in Table 4-15.

Table 4-15 Heat Recovery Ventilator

Parameter	1st Floor	2nd Floor
Availability Schedule Name	Always Available	Always Available
Nominal Supply Air Flow Rate (m³/s)	0.0205	0.0187
Sensible Effectiveness at 100% Heating Air Flow	0.72	0.72
Latent Effectiveness at 100% Heating Air Flow	0	0
Sensible Effectiveness at 75% Heating Air Flow	0.78	0.78
Latent Effectiveness at 75% Heating Air Flow	0	0
Sensible Effectiveness at 100% Cooling Air Flow	0.72	0.72
Latent Effectiveness at 100% Cooling Air Flow	0	0
Sensible Effectiveness at 75% Cooling Air Flow	0.78	0.78
Latent Effectiveness at 75% Cooling Air Flow	0	0
Nominal Electric Power	16.0	16.0
Supply Air Outlet Temperature Control	No	No
Heat Exchanger Type	Plate	Plate
Frost Control Type	Exhaust Air Recirculation	Exhaust Air Recirculation
Threshold Temperature	-5	-5
Initial Defrost Time Fraction	0.21875	0.21875
Rate of Defrost Time Fraction Increase	0.004261	0.004261
Economizer Lockout	No	No

The HRV equipment installed in the NZERTF is designed with a "boost" mode that will increase the outdoor air flow rate to 0.071 m³/s (70.8 L/s or 150 CFM) when the bathroom fans are operated. This "boost" mode is not currently included in the E+ model, which underestimates the energy use to meet the ventilation requirements. A sensitivity test was run to determine the potential impact of the boost mode by simulating the boost mode running 100 % of the time. The change in annual energy consumption relative to the baseline model is estimated to be 652 kWh. Given that the boost mode is only used when the bathrooms are in-use in the master bathroom

30

during the demonstration year (6.5 hours per week or 3.9 % of the year), the maximum impact on energy use would be 25 kWh or an increase of 0.2 %.

Figure 4-7 shows that $E+$ can allow an HRV to have bypass dampers to stop heat transfer when it is not beneficial to do so. For example, if the outdoor air temperature is lower than the HVAC exhaust air temperature in the summer and the HRV continues exchanging heat, the heat pump would need to cool the outdoor air that is being warmed by the stale exhaust air, thus wasting energy. However, these bypass dampers are not currently installed in the NZERTF.

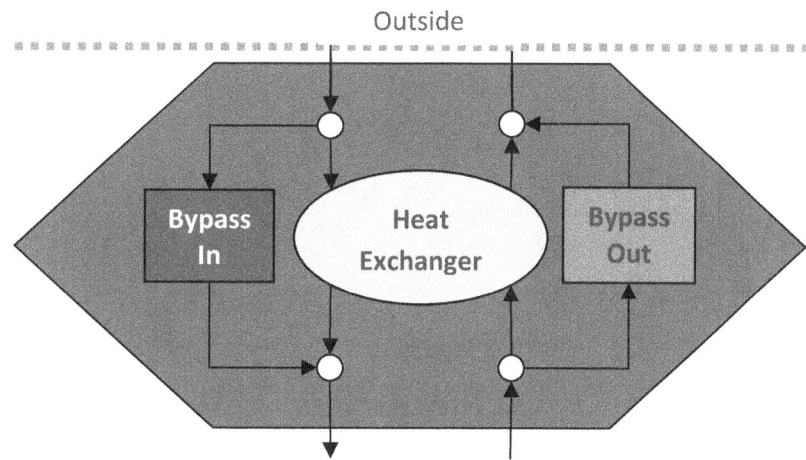

Figure 4-7 Heat Recovery Ventilator Layout

4.4.4 Humidity

The "tight" building envelope design could lead to high humidity issues throughout the year. The advanced technology heat pump being installed in the NZERTF has the capability to run in dehumidification-only mode. However, $E+$ cannot currently model such advanced equipment. The building specifications include an Ultra-Aire 70H whole house ventilating dehumidifier. However, $E+$ can only model dehumidifiers for a single zone. The humidity level in the simulation model is controlled by two DX dehumidifiers, one for each floor.

The dehumidifiers are operated based on the dehumidifying setpoint of 60 %, which is based on Hendron and Engebrecht (2010). The equipment is available year-round to run whenever the relative humidity reaches 60 % in its zone (1st floor or 2nd floor) regardless of whether the heat pump is running to meet the setpoint temperature and when the heat pump is not running when the setpoint temperature is met.

The specifications for this dehumidifier along with estimated water removal and energy factor curves from the Ultra-Aire 70H are used to define the parameters and performance curves shown in Table 4-16 and Table 4-17, which are used for the DX humidifiers in the simulation. The rated energy factor is assumed to be half (1.0 L (0.26 gal.) per kWh) of the Ultra-Aire equipment rating (2.0 L (0.53 gal.) per kWh) to ensure a conservative (high) electricity consumption estimate. The water removal rate of 30.75 L (8.1 gal.) per day for the Ultra-Aire 70H is split between the 1st floor (16.08 L [4.24 gal.] per day) and 2nd floor (14.67 L [3.86 gal.] per day) based on occupied volume. The model assumes 100 % of compressor heat is rejected into the conditioned zone. The Ultra-Aire 70H dehumidifier will be located in the basement, causing the simulation model to slightly overestimate the temperature level in each occupied zone.

Table 4-16 Dehumidifier Parameters

Field	Units	1st Floor	2nd Floor
Availability		Always Available	Always Available
Rated Water Removal	L/day (pints/day)	16.08 (33.98)	14.67 (31.00)
Rated Energy Factor	L/kWh	1.0	1.0
Rate Air Flow Rate	m^3/s (CFM)	0.897 (190)	0.897 (190)
Min. Dry-Bulb	°C (°F)	-1.1 (30.0)	-1.1 (30.0)
Max. Dry-Bulb	°C (°F)	32.2 (90.0)	32.2 (90.0)
Off-Cycle Parasitic Elect. Load	W	0.0	0.0

The dehumidifier water removal curve and energy factor curve coefficient values are shown in Table 4-17.

Table 4-17 Dehumidifier Performance Curves

Curves	Water Removal	Energy Factor
Constant	-1.281357458	-2.743752887
X	0.032064893	0.114491512
X^2	-0.000280794	-0.001456831
Y	0.028356002	0.053860412
Y^2	-0.000134939	-0.000244965
X*Y	0.000271496	-0.000362021
Min. X	4.4	4.4
Max. X	50	50
Min. Y	0	0
Max. Y	100	100
X= Dry-Bulb Temperature		

4.5 Domestic Hot Water

The domestic hot water (DHW) system installed in the NZERTF includes a number of potential combinations of equipment, including 4 solar thermal collectors, two storage tanks, a heat exchanger, and a heat pump water heater. The remainder of this section will define the DHW system and the DHW consumption simulated during the demonstration phase of the NZERTF.

4.5.1 Domestic Hot Water Heater Equipment

The DHW system simulated in the model, as shown in Figure 4-8, has a two tank system with two solar thermal panels located on the west half of the front porch and an air-to-water heat pump with a two tank system located in the basement. The solar thermal system uses a 50/50 water/glycol mix and indirectly heats the water in the storage tank through a heat exchanger. The heat pump draws water from the storage tank and will further heat the water if necessary to meet the target exit temperature of 48.9 °C (120 °F) for hot water use.

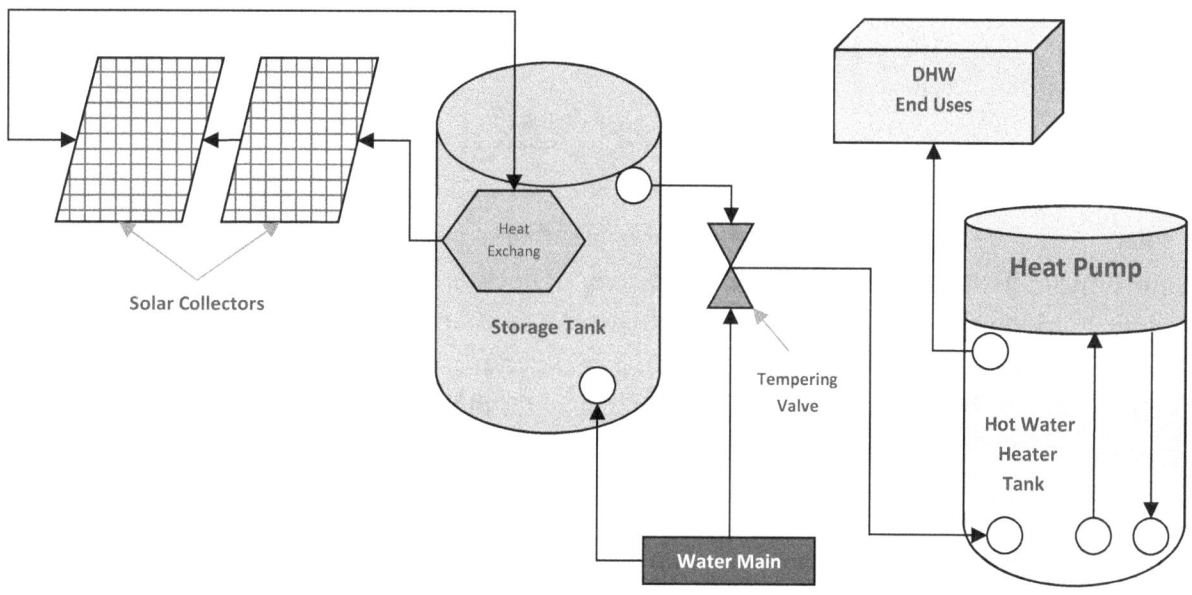

Figure 4-8 Domestic Hot Water Heating System

There are two tanks in the system, a 0.302833 m³ (80 gal.) storage tank pre-heated by the solar thermal system and a 0.189271 m³ (50 gal.) tank connected to the heat pump. The maximum temperature allowed is 76.6 °C (170.0 °F) for the storage tank and 71.1 °C (160.0 °C) for the hot water heater tank. The heat pump turns on when the water heater tank water temperature drops below 49.0 °C (120.2 °F) and turns off once the temperature increases to 54.0 °C (129.2 °F). The back-up supplemental electric heating coil will turn on if the heat pump cannot maintain the water temperature above 48.9 °C (120.2 °F). Based on the heat pump characteristics and the location, the back-up electric heater should rarely be required to meet the hot water demand. The Heliodyne HPAK heat exchanger is able to transfer 80 % of the energy from the solar thermal close-loop system.

The solar thermal collectors are Heliodyne GOBI 406 001 flat plate collectors with the performance characteristics in Table 4-18. The maximum flow rate is assumed to be the test flow rate.

Table 4-18 Solar Thermal Collector Parameters

Field	Units	Value
Gross Area	m^2	2.50281
Test Fluid		Water
Test Flow Rate	m^3/s	0.0000498425
Test Correlation Type		Inlet
Efficiency Equation Coefficient 1		0.732
Efficiency Equation Coefficient 2	W/m^2-K	-4.1949
Efficiency Equation Coefficient 3	W/m^2-K^2	0
Incident Angle Modifier Coefficient 2		0.0581
Incident Angle Modifier Coefficient 2		-0.2744
Max Flow Rate	m^3/s	0.0000498425

The solar thermal loop pump operation control is based on 4 conditions. The solar collector loop pump is turned on when the temperature in the solar collectors reaches 100 °C (212 °F). The solar collector loop pump is turned off when the temperature in the water heater reaches 80 °C (176 °F). The solar collector loop pump is turned on whenever the temperature of the fluid in the solar collector loop is 10 °C (18 °F) greater than the water in the storage tank. The solar collector loop pump is turned off when the temperature of the fluid in the solar collector loop is less than 2 °C (3.6 °F) greater than the water in the storage tank.

Two plant loops are required in the simulation: the solar collector loop and domestic hot water loop. The solar collector loop has a maximum temperature in the loop of 163 °C (325.4 °F) and a minimum temperature of -45.6 °C (-50.1 °F). The domestic hot water loop has a maximum temperature in the loop of 100 °C (212 °F) and a minimum temperature of 3.0 °C (37.4 °F). The maximum flow rate for each plant loop is autosized while the minimum loop flow rate is 0.0 m^3/s (0.0 CFM). The loop is designed for an exit temperature, the temperature supplied to the "use side," is 48.9 °C (120.0 °F) for the domestic hot water loop and 100 °C (212 °F) for the solar collector loop. For the solar collector loop, it is the temperature of the water that exits the storage tank. For the domestic hot water loop, it's the temperature of the water that is exiting the heat pump.

The storage tank and the hot water heater tank are both stratified tanks with 6 nodes and have the parameter values listed in Table 4-19. The storage tank inlet and outlet for the solar collector loop are assumed to be at the highest and lowest node height, respectively. The hot water heater tank parameters are based on the specifications for the installed equipment in the NZERTF.

Table 4-19 Storage Tank and Hot Water Heater Tank Parameters

Field	Units	Storage Tank	Hot Water
Tank Volume	m3	0.302833	0.189271
Tank Height		1.59385	1.143
Tank Shape		Vertical Cylinder	Vertical Cylinder
Max. Temp. Limit		170	160
Heater 1 Setpoint Temperature	°C		48.9
Heater 1 Deadband Temperature Difference	Δ°C		5.0
Heater 1 Maximum Capacity	W		3800
Heater 1 Height	M		0.86
Heater 2 Setpoint Temperature	°C		48.9
Heater 2 Deadband Temperature Difference	Δ°C		10.0
Heater 2 Maximum Capacity	W		3800
Heater 2 Height	M		0.5
Ambient Temperature Indicator		Zone	Zone
Ambient Temperature Zone Name		Basement	Basement
Uniform Skin Loss Coefficient to Ambient Temperature	W/m^2-K	0.846	0.41
Skin Loss Fraction to Zone		1.0	1.0
Off Cycle Flue Loss Coefficient to Ambient Temperature	W/K	0.0	0.0
Off Cycle Flue Loss Fraction to Zone		1.0	1.0
Use Side Effectiveness		1.0	1.0
Use Side Inlet Height	m	0.398463	0.2286
Use Side Outlet Height	m	1.46103	1.0922
Source Side Effectiveness		0.80	1.0
Source Side Inlet Height	m	1.13665	0.2286
Source Side Outlet Height	m	0.398463	0.2286
Inlet Mode		Fixed	Fixed
Use Side Design Flow Rate	m^3/s	Autosize	Autosize
Source Side Design Flow Rate	m^3/s	Autosize	Autosize
Indirect Water Heating Recovery Time	hr	1.5	1.5
Number of Nodes		6	6

The uniform skin loss coefficient to the ambient temperature in W/m^2-K must be calculated based on the standby heat loss from the storage tank of 0.2 °C/hr (0.36 °F/hr) using the following two step approach. First, the rate of energy loss in standby mode (Q) must be calculated using the following equation.

$$Q = V * \rho * c_p * \Delta T$$

Where $V = volume\ of\ water = 45\ gal$

$\rho = density\ of\ water = 8.217\ ^{lb}/_{gal}$

$c_p = specific\ heat\ of\ water = 1\ ^{Btu}/_{°F\ \cdot\ lb}$

$$\Delta T = rate\ of\ temperature\ decrease = 0.36\ {}^{\circ}F/_{hr}$$

Given these values, the energy loss rate in standby mode (Q) is 133 Btu/h, which converts to 39 W. The second step is to calculate the uniform skin loss coefficient to the ambient temperature (USL), which is calculated using the SI value of Q and the following equation.

$$USL = \frac{Q}{A * (T_{Tank} - T_{Ambient})}$$

Where $USL = uniform\ skin\ loss\ coeffient$
  $A = tank\ surface\ area = \pi * diameter_{Tank} * height = 2.53\ m^2$
  $T_{Tank} = temperature\ of\ water\ in\ tank = 330.4\ K\ (135\ {}^{\circ}F)$
  $T_{Ambient} = temperature\ of\ ambient\ air = 292.9\ K\ (67.5\ {}^{\circ}F)$

Given these values, the uniform skin loss coefficient (USL) is 0.41 W/m²-K.

The air-to-water hot water heat pump is a Hubbell PBX 50-SL, and has the operation and performance parameter values listed in Table 4-20 and Table 4-21, respectively. The heat pump operation is based on the water temperature at the height of the second electric heating element (0.5 m).

Table 4-20 Hot Water Heat Pump Operation Parameters

Field	Units	Value
Dead band Temperature Difference	Δ°C	5.0
Compressor Set point Temperature	°C	53.9
Condenser Water Flow Rate	m³/s	Autocalculate
Evaporator Air Flow Rate	m³/s	Autocalculate
Inlet Air Configuration		Zone Air Only
Inlet Air Zone Name		Basement
Min. Inlet Air Temp. for Compressor Operation	°C	5
Compressor Location		Basement
Parasitic Heat Rejection		Basement
Fan Placement		Draw Through
Temperature Control Sensor Location		Heater 2

The heating coil for the air-to-water heat pump has a rated capacity of 1375 W and a COP of 2.6. A factor not accounted for in the simulation model is that the heat pump will slightly dehumidify the basement while operating.

Table 4-21 Hot Water Heat Pump Coil Parameters

Field	Units	Value
Rated Capacity	W	1375
Rated COP	W/W	2.6
Rated Sensible Heat Ratio		0.85
Rated Evaporator Inlet Air DB Temp.	°C	19.7
Rated Evaporator Inlet Air WB Temp.	°C	13.5
Rated Condenser Inlet Water Temp.	°C	57.5
Rated Evaporator Air Flow Rate		Autocalculate
Rated Condenser Air Flow Rate		Autocalculate
Evaporator Fan Power Included in Rated COP		Yes
Condenser Pump Power Included in Rated COP		No
Condenser Pump Heat Included in Heat Cap. And COP		No
Condenser Water Pump Power	W	0
Fraction of Condenser Pump Heat to Water		0.2

The hot water heat pump fan and DHW fan parameter values are listed in Table 4-22.

Table 4-22 Hot Water Heat Pump Fan and DHW Fan Parameters

Field	Fans
Availability	Always Available
Fan Efficiency	80 %
Pressure Rise	100 Pa
Maximum Flow Rate	Autosize
Motor Efficiency	90 %
Motor in Airstream Fraction	1.0

There are two intermittent pumps used in the DHW system, one for the solar thermal collectors and one for the DHW loop. The parameter values for both pumps are shown in Table 4-23.

Table 4-23 Domestic Hot Water Loop Pump Parameters

Field	Units	DHW	Solar Collector
Rate Flow Rate	m³/s	autosize	0.000099685
Rated Pump Head	Pa	15000	15000
Rated Power Consumption	W	Autosize	Autosize
Motor Efficiency		0.87	0.87
Fraction of Motor Inefficiency to Fluid Stream		0	0
Part Load Pref. Curve – Coefficient 1		0	0
Part Load Pref. Curve – Coefficient 2		1	1
Part Load Pref. Curve – Coefficient 3		0	0
Part Load Pref. Curve – Coefficient 4		0	0
Minimum Flow Rate	m³/s	0	0
Pump control Type		Intermittent	Intermittent

4.5.2 Domestic Hot Water End Use

The DHW end use in the NZERTF has 5 categories: bath, shower, sinks, dishwasher, and clothes washer. The use schedules in terms of fraction of peak flow (FPF) are shown in Figure 4-9, and are based on Omar (Forthcoming). The clothes washer is used 5 times a week, one load on Wednesday and two loads on both Saturday and Sunday. The dishwasher is used for one load five days a week: Monday, Wednesday, Friday, Saturday, and Sunday. The sink, shower, and bath water draws vary by the occupancy schedule to match the target water use from Hendron and Engebrecht (2010).

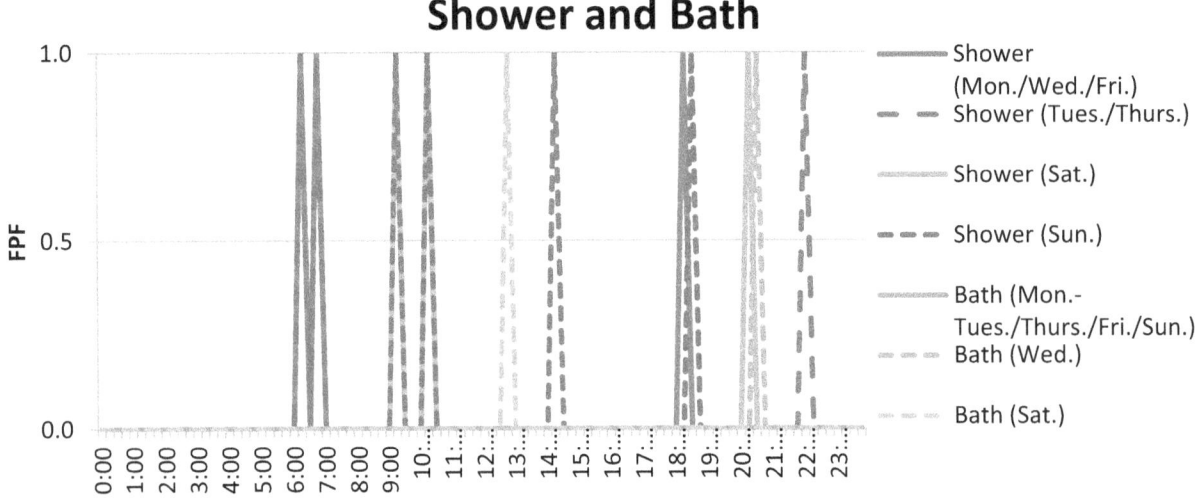

Sinks

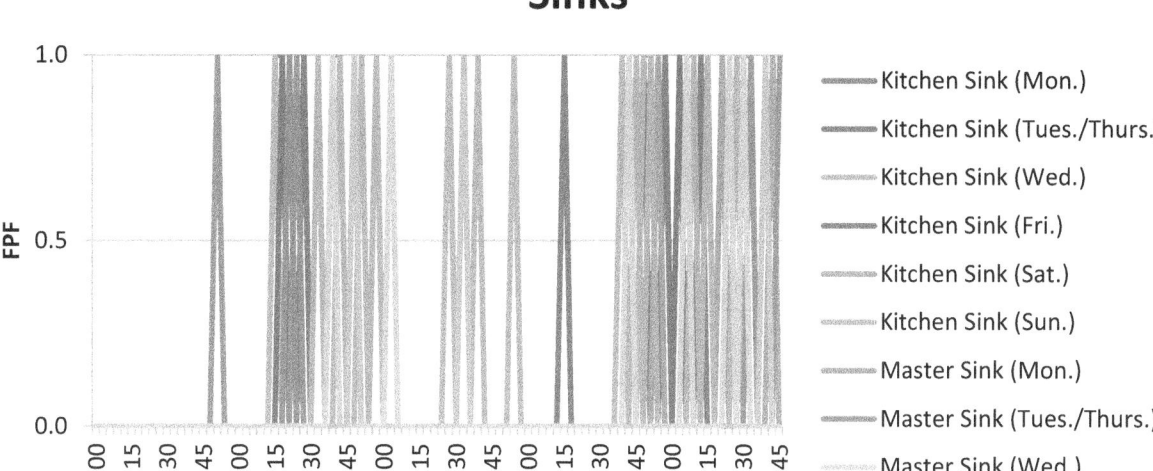

Cothes Washer and Dishwasher

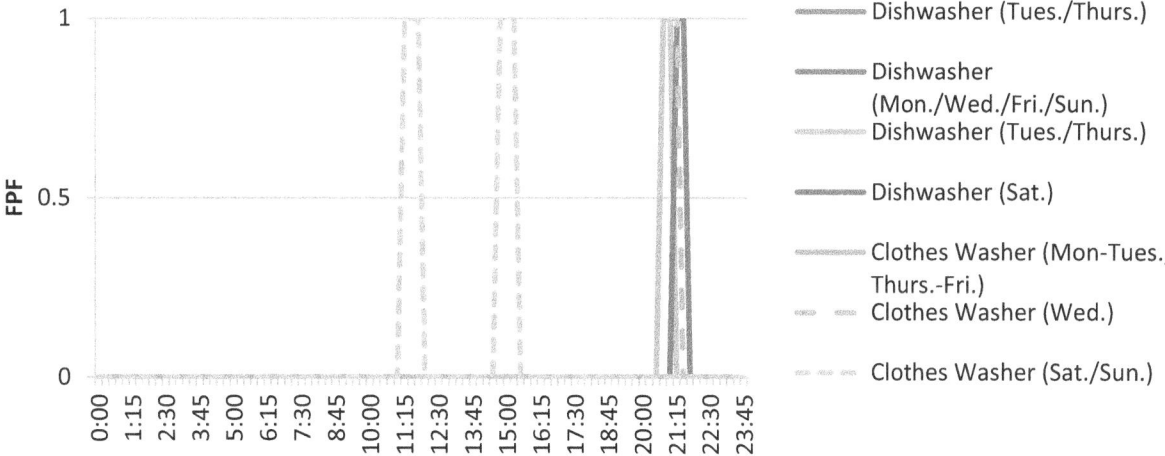

Figure 4-9 Domestic Hot Water End Use Schedules – Fraction of Peak Flow

Table 4-24 defines DHW use in terms of daily water use, water temperature, and heat gain fractions. Hendron and Engebrecht (2010) is used to calculate daily water consumption and heat gains for each of the categories except for the dishwasher and clothes washer, which are based on Omar (Forthcoming) and the equipment's Energy Star ratings.

Table 4-24 Domestic Hot Water Use and Thermal Load Fractions

DHW Use	Water Temp.	Daily	Sensible Load	Latent Load
Clothes Washer	48.9 (120.0)	35.7 (9.43)	0.0	0.0
Dishwasher	48.9 (120.0)	4.2 (1.12)	0.0	0.0
Shower	40.6 (105.0)	142.4 (37.6)	0.2127	0.005655
Bath	40.6 (105.0)	11.3 (3.0)	0.2083	0.0
Sinks	40.6 (105.0)	93.6 (24.7)	0.09869	0.00127

The sink, shower, and bath water draws are assumed to be mixed water use at 40.6 °C (105.1 °F). The sink flow rates are 5.7 L (1.5 gal.) per minute while the shower and bath flow rates are 6.6 L (1.75 gal.) per minute. The sensible and latent load fractions for the sink, shower, and bath draws are calculated by reverse engineering using the heat gain estimate equations in Hendron and Engebrecht (2010). Omar (Forthcoming) defines sink draws in 21 second intervals while the *E+* model can only model at one minute increments. The consolidation of draws results in the total length of draws for a week being 6 seconds longer than Omar (Forthcoming), or additional hot water use of 29.5 L (7.8 gal.). The additional DHW use will slightly overestimate electricity use relative to Omar (Forthcoming).

The domestic hot water use equipment is assumed to use 100 % hot water and has no sensible or latent heat gains resulting from the hot water use. The clothes washer is assumed to consume 46.4 L (12.25 gal.) per cycle. Total water use is the product of the water use per cycle and the number of loads per year (260) for a total of 12 057 L (3185 gal.) per year or 33.0 L (8.73 gal.) per day. The dishwasher is assumed to consume 5.9 L (1.57 gal.) per cycle. Total water use is the product of the water use per cycle and the number of loads per year (260) for a total of 1545 L (408 gal.) per year or (1.12 gal.) per day.

4.6 Solar Photovoltaic

Energy efficient design reduces but does not eliminate electricity use by the NZERTF. Solar photovoltaic (PV) panels are installed on the roof of the NZERTF to produce at least as much electricity as is consumed by the NZERTF on an annual basis.

Figure 3-2 shows that there are 32 SunPower SPR-320E-WHT-U solar PV panels installed on the roof of the NZERTF in 4 horizontal strings of 8 panels each. Two strings are connected to a SunPower 5000m LUT inverter for a total of 2 inverters. The solar panel parameters are listed in Table 4-25.

Table 4-25 Solar Photovoltaic Panel Parameters

Field	Units	Value
Cell Type		Crystalline Silicon
Number of Cells in Series		96
Active Area	m^2	1.472
Transmittance Absorptance Product		0.95
Semiconductor Bandgap	eV	1.12
Shunt Resistance	Ohms	1125
Short Circuit Current	A	6.24
Open Circuit Voltage	V	64.8
Reference Temperature	C	25
Reference Insolation	W/m^2	1000
Module Current at Maximum Power	A	5.86
Module Voltage at Maximum Power	V	54.7
Temp. Coeff. Of Short Circuit Current	A/K	0.0035
Temp. Coeff. Of Open Circuit Voltage	V/K	-0.1766
Nominal Operating Cell Temp. Test Ambient Temp.	C	20
Nominal Operating Cell Temp. Test Cell Temp.	C	45
Nominal Operating Cell Temp. Test Insolation	W/m^2	800
Module Heat Loss Coefficient	W/m^2-K	30.4
Total Heat Capacity	J/m^2-K	7984

The solar PV inverter parameters for the SunPower 5000m LUT are listed in Table 4-26. The inverter efficiency is greatest at 50 % its power and nominal voltage. The inverters are located in the main attic area, and the energy lost by the inverter enters the attic space.

Table 4-26 Solar Photovoltaic Inverter Parameters

Field	Units	Value
Availability Schedule Name		Always Available
Zone Name		Main Attic (Zone 4)
Radiative Fraction		0.05
Rated Maximum Continuous Output Power	W	5000
Night Tare Loss Power	W	0.1
Nominal Voltage Input	V	438
Eff. At 10 % Power and Nominal Voltage		0.935
Eff. At 20 % Power and Nominal Voltage		0.962
Eff. At 30 % Power and Nominal Voltage		0.968
Eff. At 50 % Power and Nominal Voltage		0.969
Eff. At 75 % Power and Nominal Voltage		0.964
Eff. At 100 % Power and Nominal Voltage		0.957

5 Simulation Results

There are so many different energy efficiency measures designed into the NZERTF that it's uncertain how these measures will interact and whether the house will perform at its net zero annual energy use goal. The *E+* model is used to confirm the house's design will perform at net zero.

5.1 Total Electricity Use

The annual electricity use is 10 317 kWh (36.2 kWh/m² or 3.4 kWh/ft²). This is 73 % lower than the average weather-adjusted home in the Northeast of 132.4 kWh/m² or 12.30 kWh/ft² (RECS, 2005). The results are even more significant because electricity is used instead of more efficient fuels, such as natural gas or fuel oil. In reality, the most likely heating fuel source would be natural gas, which is used by 60.3 % of all housing units in the Middle Atlantic census subregion followed by fuel oil at 25.2 % while electricity is the primary heating fuel source for only 8.6 % of households in the subregion (RECS, 2005). The breakdown of the general electricity use categories can be seen in Figure 5-1. The efficiency of the NZERTF results in HVAC operation (the combination of heating, cooling, fans, and heat recovery) accounting for 43 % of total electricity use, which is 15 percentage points lower than the average household in the Middle Atlantic census subregion (RECS, 2005).

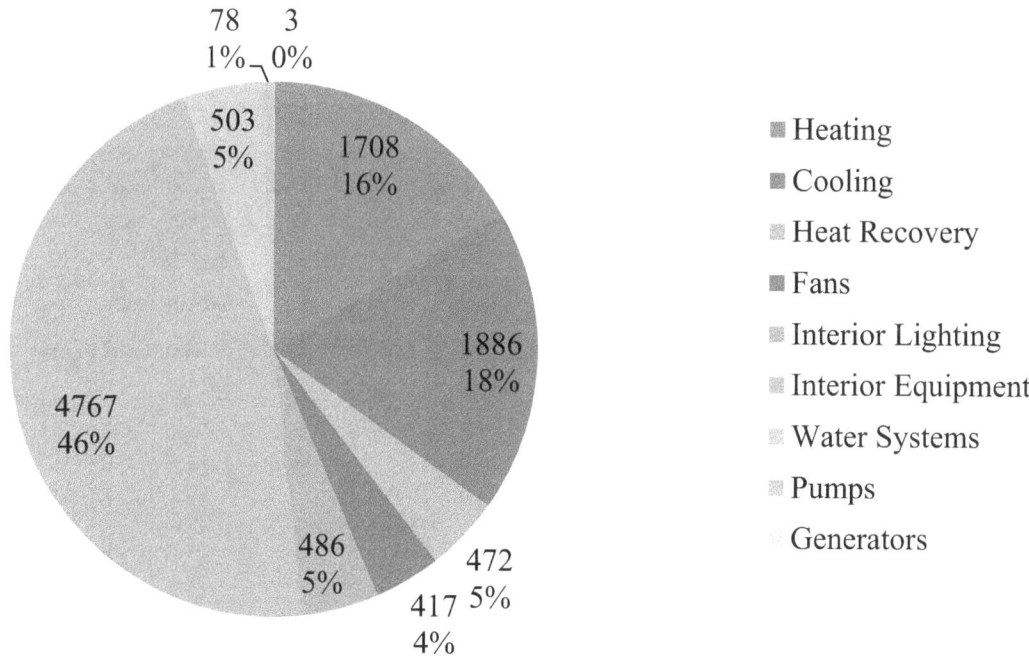

Figure 5-1 Electricity Use by Category - kWh and Percentage

HVAC (heating and cooling) and DHW heating loads are the only electricity loads that vary significantly across months. So it is not surprising that Figure 5-2 shows that the most electricity use occurs during the coldest and warmest months (January, February, December, July, and August).

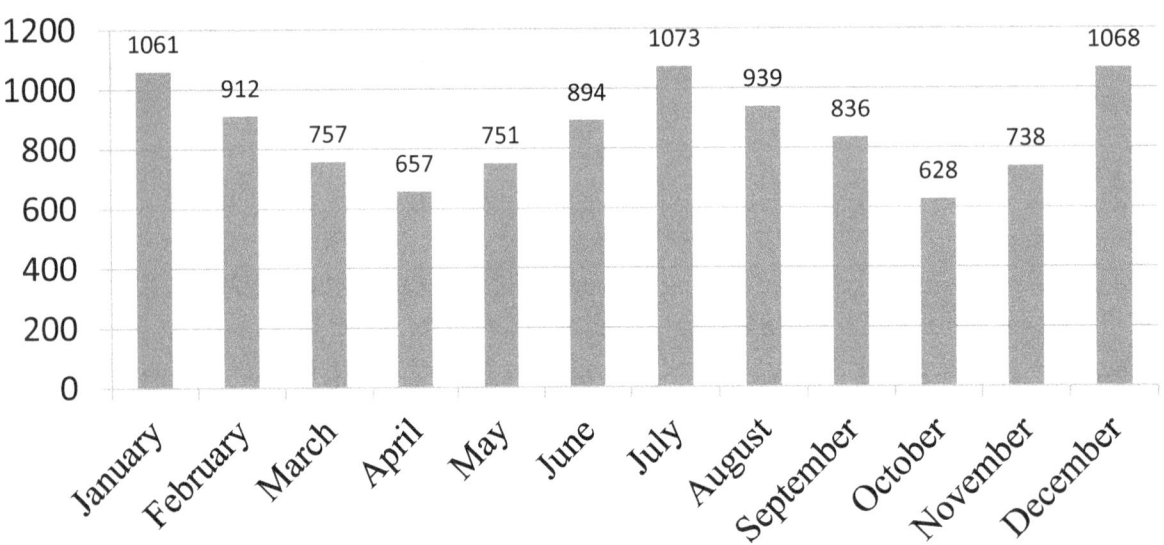

Figure 5-2 Total Electricity Use by Month - kWh

5.2 Lighting

Figure 5-3 shows that lighting loads are greatest on the 1st floor of the house, which accounts for 78 % of total electricity consumption from lighting. Nearly half of total lighting loads occur in the kitchen (48 %) and over a quarter occur in the living room (28 %).

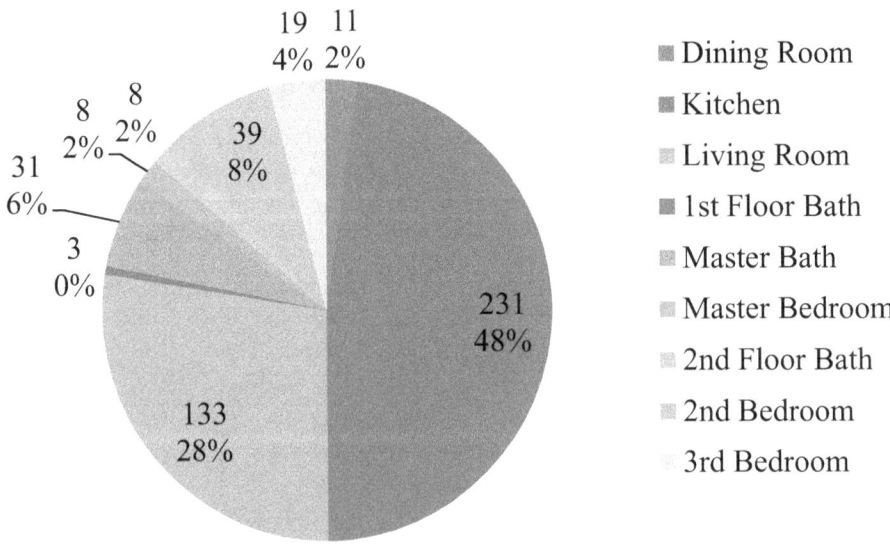

Figure 5-3 Lighting Electricity Use by Room – kWh and Percentage

5.3   Interior Equipment

The category with the greatest amount of electricity use (41 %) is interior equipment, which can be split into the 16 subcategories in Figure 5-4. Large appliances are separated out while miscellaneous electric loads (MELs) are grouped by room. Large appliances account for 51 % of electricity use by interior equipment. The clothes washer and clothes dryer account for 50 % of all non-MEL electrical loads while the cooking equipment (oven, cooktop, microwave, and range hood) account for 647 kWh of electricity consumption. MELs include televisions, cable boxes, hair dryers, portable electric devices, etc. The room with the greatest electricity consumption is the office (703 kWh), which is driven by the desktop computer, phone, and printer. There are large MEL loads in both the living room (567 kWh) and the master bedroom (511 kWh) because those are the rooms in which televisions, Blu-Ray players, and cable boxes are located.

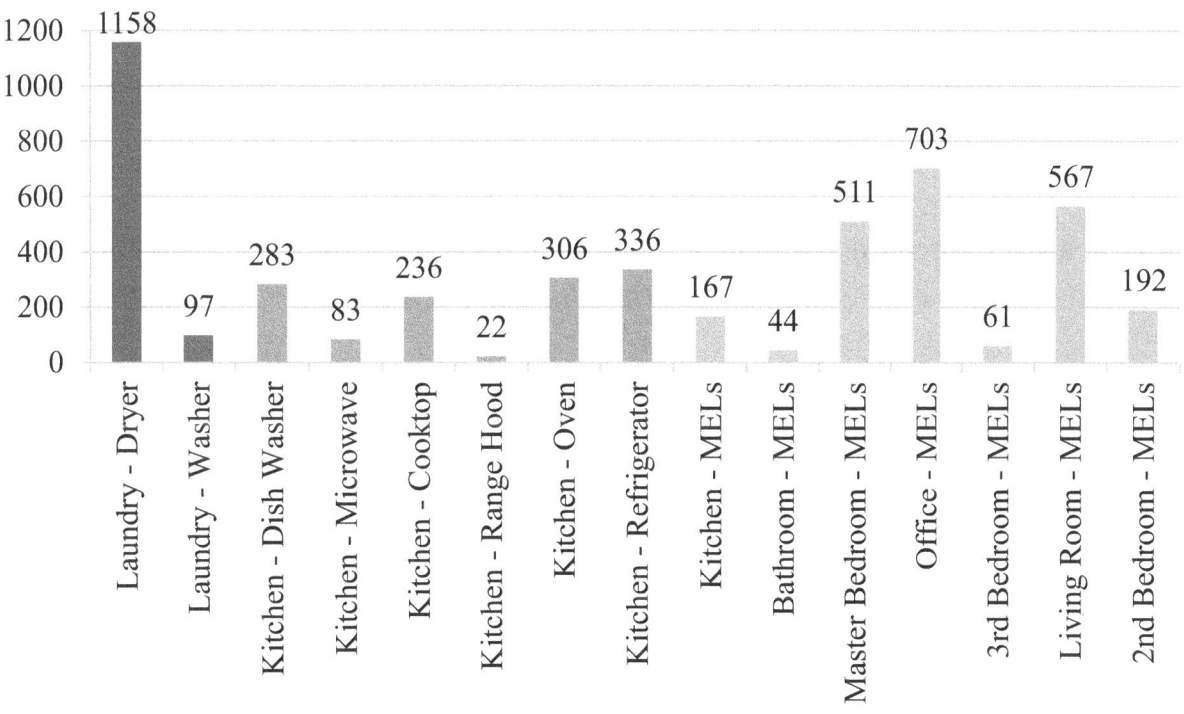

Figure 5-4 Equipment Electricity Use - kWh

5.4   HVAC

The electricity use to meet the HVAC load, shown in Figure 5-5, is split into 4 subcategories. Heating accounts for 38 % of electricity used by the HVAC system while 42 % is used for cooling with the remainder used for the heat pump fan (7 %), HRV operation (11 %), and HRV fans (2 %).

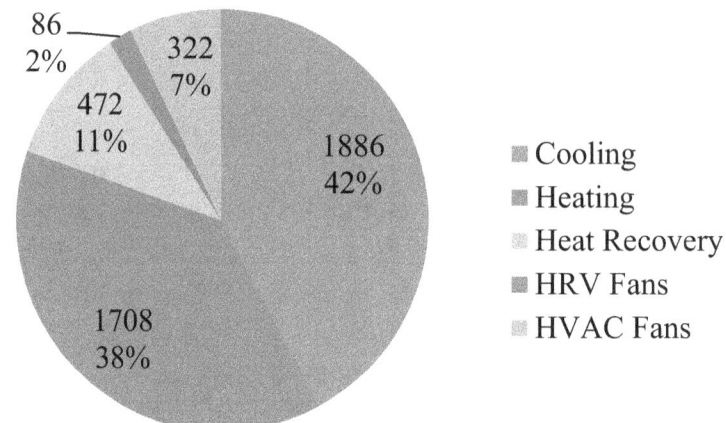

Figure 5-5 HVAC Electricity Use – kWh and Percentage

Cooling electricity use is 4 % greater than heating electricity use even though the number of heating degree days for the location is over 4 times greater than the cooling degree days. There are two key reasons for this discrepancy: (1) building envelope is very "tight" and (2) heat recovery benefits are greater for heating than for cooling. The tight design of the NZERTF exterior envelope keeps the internal gains from occupants, equipment, and DHW use within the NZERTF, making it easier to heat and harder to cool the NZERTF. The heat recovery system is more beneficial when there is a greater difference between the indoor and outdoor temperatures. Figure 5-6 shows more hours for which the outdoor dry-bulb (ODB) temperature is lower than the indoor temperature as well as greater energy transfer as the ODB decreases.

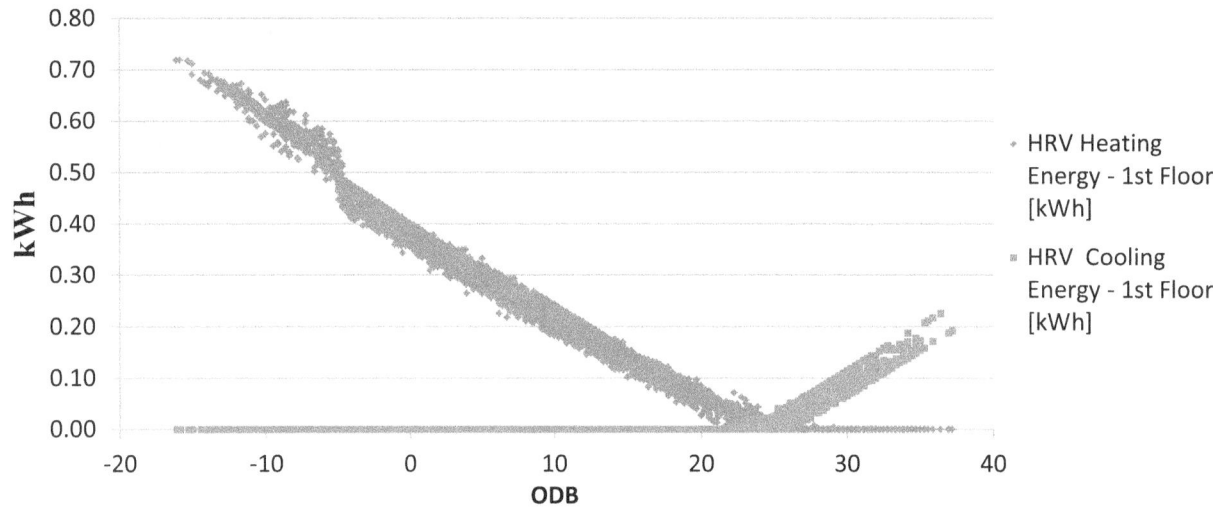

Figure 5-6 HRV Hourly Energy Transfer by ODB - kWh[13]

---

[13] Data is based on a 1 hour time step because it shows the desired pattern, but requires less data than a 1 minute time step.

Figure 5-7 shows that the energy transferred by the heat exchanger is greatest for December through February, ranging from 467 kWh (1681 MJ) to 546 kWh (1967 MJ) per month. March, April, October, and November range from 252 kWh (908 MJ) to 364 kWh (1311 MJ). The HRV transfers the least amount of energy during cooling season, where May through September ranges from 80 kWh (288 MJ) to 174 kWh (625 MJ) per month. In some cases, the HRV actually works against the HVAC system during cooling season because the air entering the house is actually cooler than the air exiting the house.

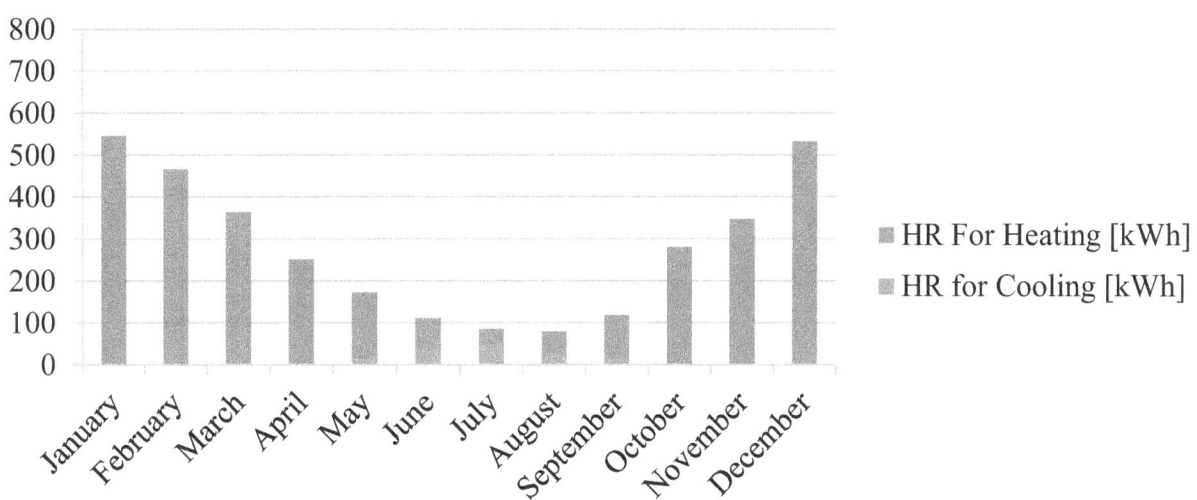

Figure 5-7 HVAC Heat Exchanger Energy Transfer (kWh) - Monthly

The monthly cooling and heating loads can be seen in Figure 5-8 and Figure 5-9. The entire cooling load is met by the heat pump while heating uses the heat pump and a supplemental electric resistance heater. Cooling only occurs April through October with July requiring the greatest cooling load.

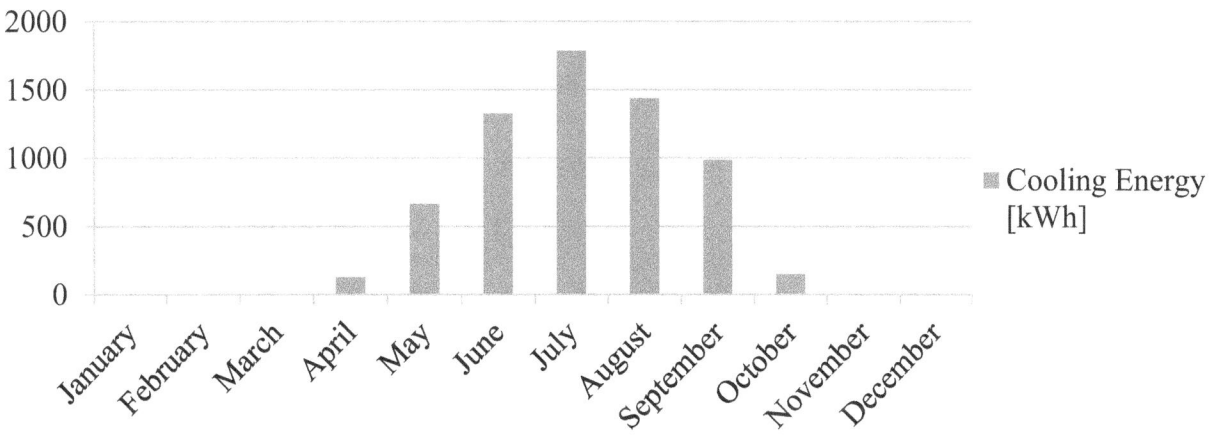

Figure 5-8 HVAC Cooling Energy (kWh) - Monthly

47

As can be seen in Figure 5-9, supplemental heating is only used briefly in January, February, and December while it is never used throughout the remainder of the heating season. The supplemental heating tends to be used during the coldest hours of the year. All supplemental heating occurs at an ODB temperature less than -10° C (14° F). No heating is required for months May through September.

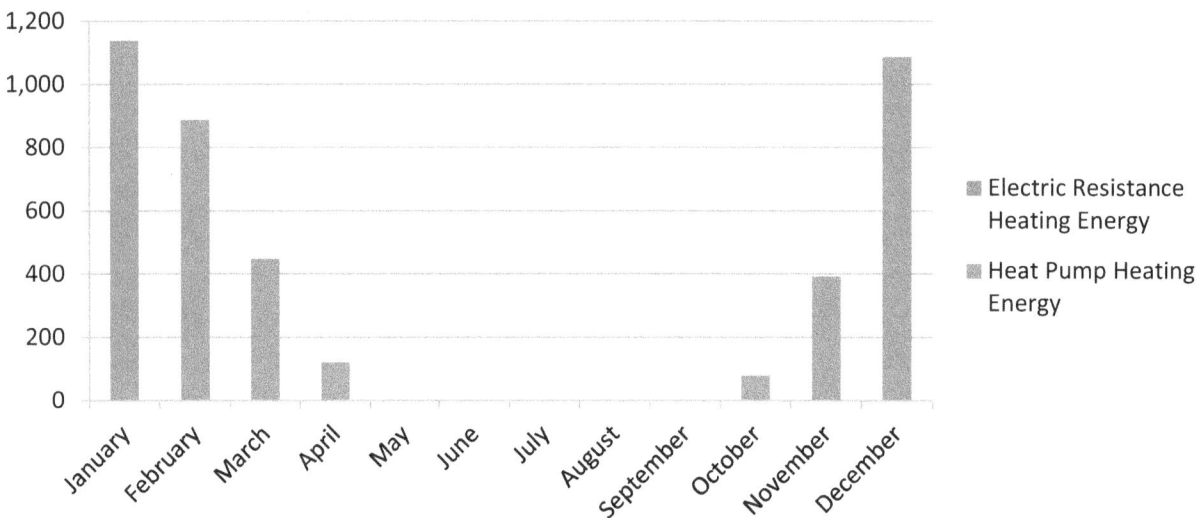

Figure 5-9 HVAC Heating Energy (kWh) - Monthly

The total loads are greater for cooling (61 %) than for heating (39 %) due to the reasons discussed at the beginning of Section 5.4. As shown in Figure 5-10, cooling loads reach 1788 kWh (6437 MJ) in July while heating loads reach 1138 kWh (4097 MJ) in January.

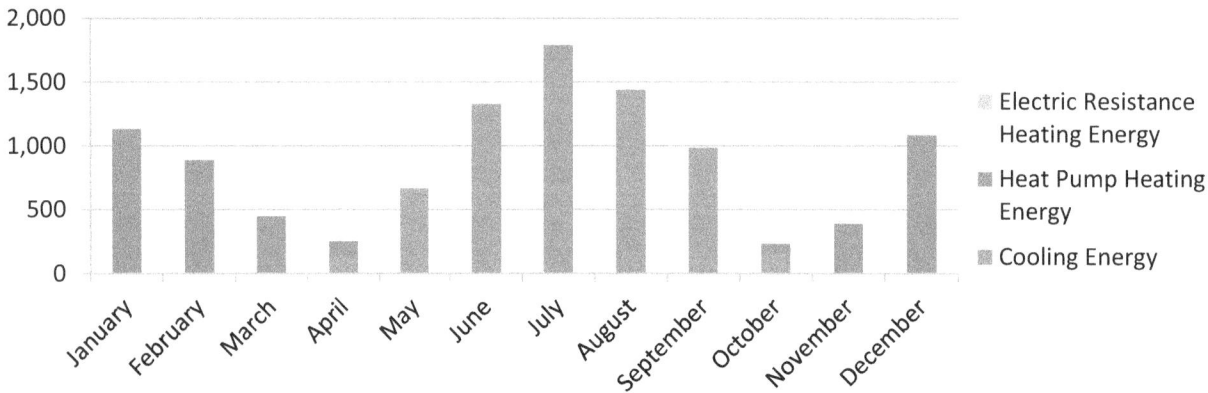

Figure 5-10 HVAC Energy Load by Coil (kWh) - Monthly

Figure 5-11 shows the electricity consumption to meet the monthly heating and cooling loads. It is odd that there is some cooling electricity consumption in nearly every month even though the DX cooling coil does not transfer any energy during the winter months.

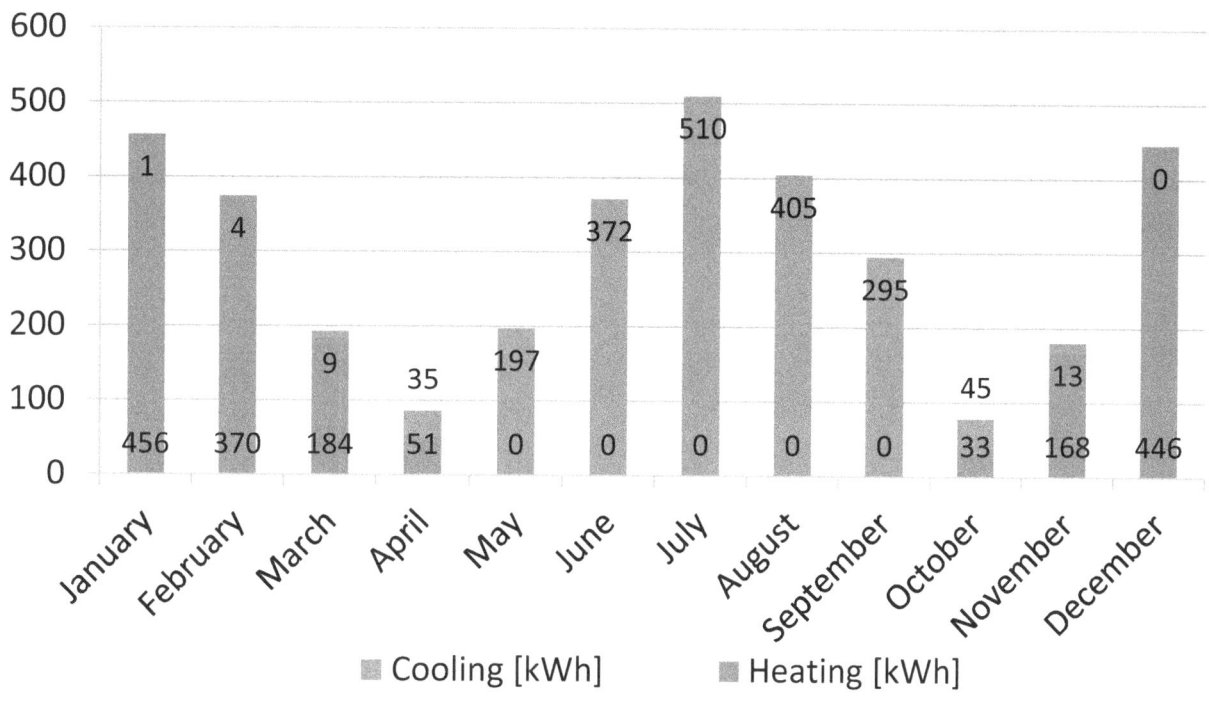

Figure 5-11 HVAC Electricity Use (kWh) – Monthly

Figure 5-12 shows the monthly electricity use for the 1st floor and 2nd floor dehumidifiers, which are highly correlated. As would be expected, the greatest amount of dehumidification is required during the summer months. Some dehumidification is necessary to keep the humidity level below 60 % in January, February, March, October, and November, which is driven by the "tight" building envelope keeping the internal latent heat gains within the NZERTF during mild temperature months.

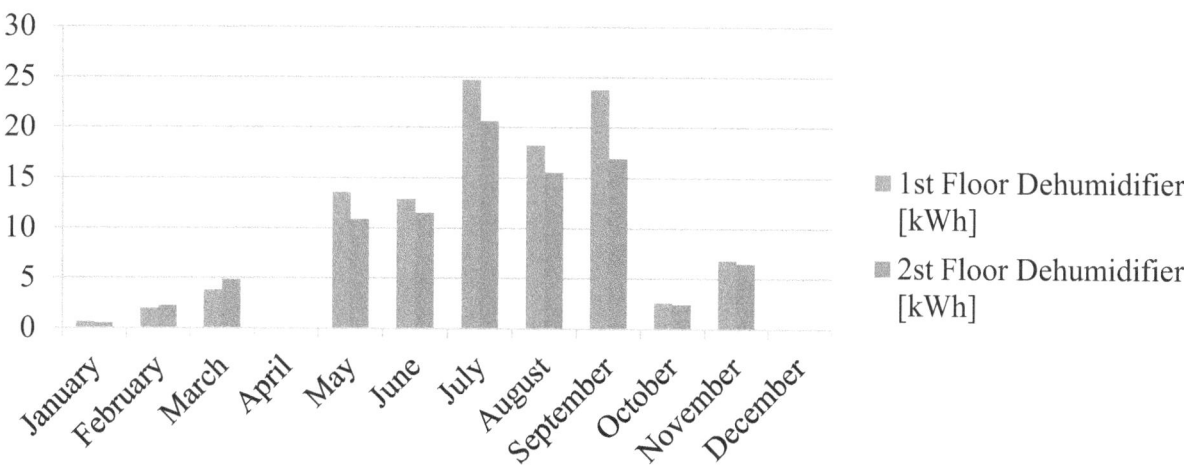

Figure 5-12 Dehumidifier Electricity Consumption by Floor – kWh

5.5 Internal Heat Gains

Figure 5-13 and Figure 5-14 show that the internal gains for both the 1st floor and 2nd floor of the NZERTF do not fluctuate significantly month-by-month, largely driven by the number of days in a month. The 1st floor heat gains in Figure 5-13 are driven by the electrical equipment loads because all the large appliances are located on that floor.

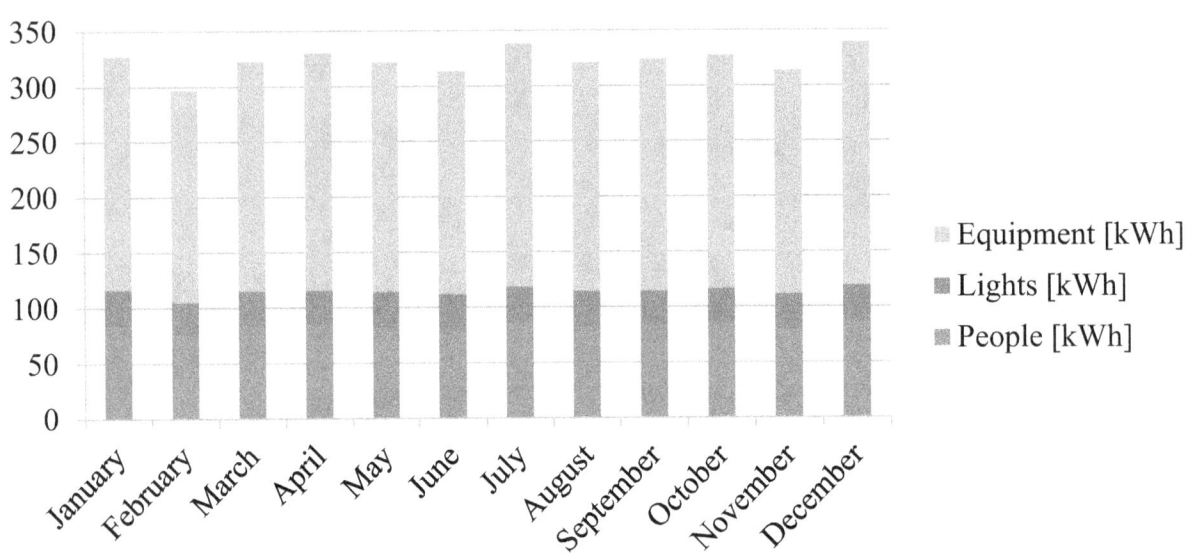

Figure 5-13 Internal Heat Gains by Category (kWh) – 1st Floor

The 2nd floor loads in Figure 5-14 are driven by the occupant loads. Lighting results in a small amount of internal gains because of the efficient light bulbs used in the house (CFL and LED).

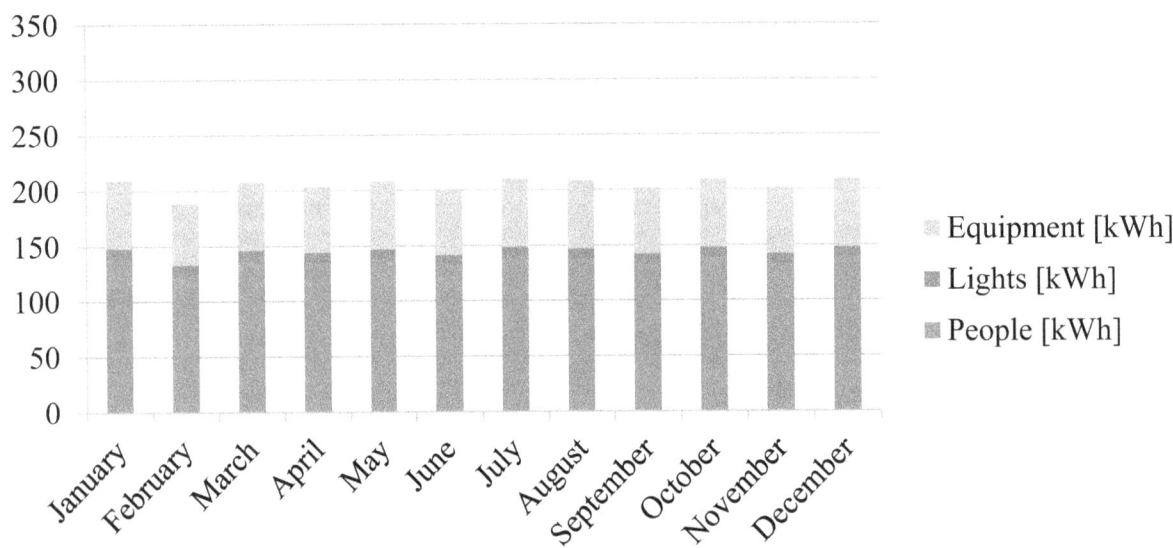

Figure 5-14 Internal Heat Gains by Category (kWh) – 2nd Floor

5.6   Solar Photovoltaic Generation

The *E+* model estimates annual solar PV production of 15 471 kWh, which is 150 % of the house's total electricity use (10 317 kWh). Given the assumptions used in the current simulation, the NZERTF not only reaches, but surpasses well beyond its net zero energy goal.[14] shows that the solar PV electricity production varies by month. As would be expected, the summer months are when the most power is produced while the winter months are when production lags. However, even with the varying monthly production, there are only 3 months (January, February, and December) for which the solar PV does not produce more electricity than is consumed by the NZERTF.

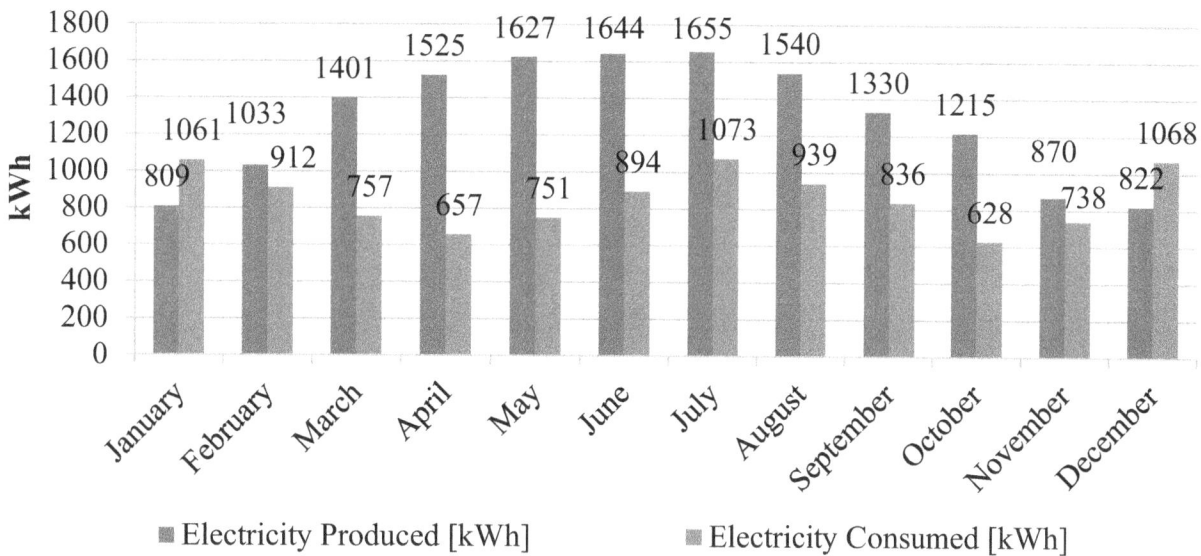

Figure 5-15 Total Electricity Consumption and Solar PV Production (kWh) - Monthly

Alternative solar PV estimates, both through different parameter values and an alternative PV software model, are used to test the sensitivity of the electricity production of the NZERTF. Two alternative *E+* solar PV modeling approaches are used to estimate electricity production. Assuming the "simple" solar PV modeling approach with a 19.6 % PV efficiency results in an estimate of 14 650 kWh (142 %). A more conservative 18.0 % PV efficiency further reduces electricity production to 13 500 kWh (131 %). Additionally, a simplified PV model (PVWATTS) was used by a NIST solar PV expert, and based on the 18.4 degree tilt of the NZERTF solar PV panels, the predicted production is 12 177 kWh (118 %). The actual efficiency performance of the solar PV panels will make a significant difference in the energy performance of the NZERTF. However, under all four approaches the NZERTF reaches its net-zero energy goal.

---

[14] This is driven by the assumed solar PV performance, which can vary significantly depending on its operating conditions.

Even assuming the most conservative estimate, the home owners are able to sell a net 1860 kWh back to the electric utility ($292 annually at 15.69¢/kWh)[15]. Putting that in context, 4 of the 32 solar photovoltaic panels could be removed from the roof and the NZERTF would still consume less electricity than it produced on an annual basis. An alternative use of the excess electricity would be to power an electric car for 5472 miles.[16]

5.7 Coil Capacities

The HVAC system includes 3 coils, a cooling coil, heating coil, and electric resistance supplemental heating coil. The $E+$ model assumes the primary cooling coil and heating coils are both multispeed DX coils with the characteristics in Table 5-1. There is a supplemental electric resistance heating coil to assist the heat pump either when the heat pump cannot meet the load or the outdoor conditions do not allow heat pump operation. The characteristics of each coil are based on the equipment installed in the NZERTF.

Table 5-1 HVAC System Coil Characteristics

HVAC System Coils	Speed	Size W (Tons)	Nominal Efficiency (W/W)
DX Cooling Coil	High	7751 (2.2)	3.69
	Low	5483 (1.6)	3.73
DX Heating Coil	High	7675 (2.2)	4.19
	Low	4908 (1.4)	4.02
Supplemental Electric Resistance	NA	3825 (1.1)	1.00

The domestic hot water system uses a hot water heat pump and supplemental electric resistance heat with the characteristics shown in Table 5-2. The characteristics for the heat pump and supplemental heating coils are based on the equipment installed in the NZERTF.

Table 5-2 DHW Heat Pump Coil Characteristics

DHW Coils	Capacity W (Tons)	Efficiency (W/W)
Hot Water Heat Pump	1375 (0.4)	2.60
Electric Resistance Heating Elements (2)	3800 (1.1)	0.98

5.8 Domestic Hot Water Electricity Use

The DHW system uses electricity to run the heat pump water heater and solar thermal water pump. Although the hot water heater has a supplemental electric resistance heater available, the electric heater is rarely required (44 kWh) due to the combination of an efficient heat pump and

---

[15] The average price per kWh of electricity is based on data from the EIA (2011) for PEPCO during 2010.
[16] Assumes an electric car gets 2.94 miles per kWh (EPA, 2012).

the pre-heating of water by the solar thermal system. Figure 5-16 shows that electricity used to meet DHW demand varies by month, and is driven solely by the heat pump water heater (458 kWh) because the water pumps require minimal electricity (81 kWh annually). Electricity use to meet DHW demand trends down as the MAT increases and then begins to rise as the MAT decreases. The solar thermal system transfers more energy to the storage tank, which lowers the amount of energy needed from the heat pump to reach the target temperature for DHW use.

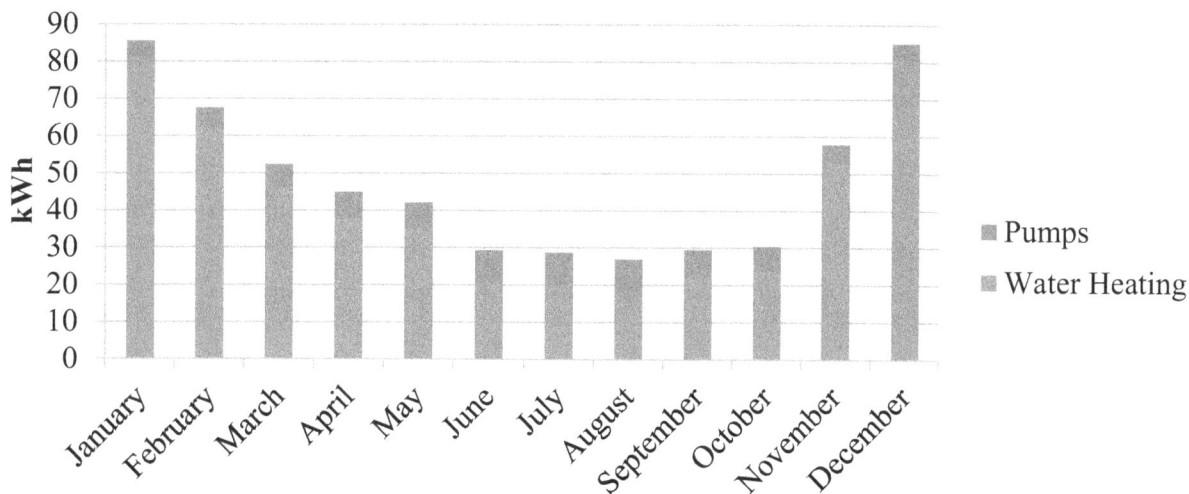

Figure 5-16 Heat Pump Water Heater and Pump Electricity Use (kWh) – Monthly

5.9   Domestic Hot Water Use

Figure 5-17 shows total DHW use is 103 970 L (27 466 gal.), or 284.8 L (75.2 gal.) per day: 54 % of which is used in baths and showers, 33 % in sinks, 12 % by the clothes washer, and 1 % for the dishwasher. Showers, baths, and sinks use a mix of hot and cold water while the washers use only hot water. The hot water use by the clothes washer would decrease if the occupants were to use cold water for a portion of their loads of laundry.

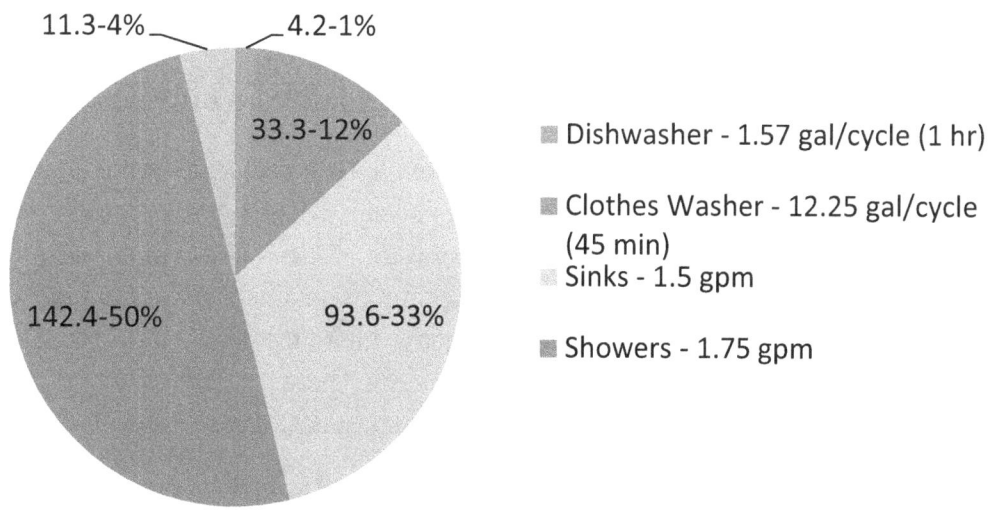

Figure 5-17 DHW Use by L/day and Percentage

The solar thermal system generates 2575 kWh (9271 MJ) of energy with the first solar collector in the series generating slightly more energy transfer. Figure 5-18 shows that total energy production by the solar thermal system is greatest in the summer while the heat pump and electric resistance supplemental heating coils transfer more energy during the winter months. The heat pump uses 458 kWh (1649 MJ) of electricity and has an efficiency of 2.6, which means it adds approximately 1191 kWh (4288 MJ) of energy to the hot water over the year. The electric resistance supplemental heating coils consume 44 kWh (158 MJ) with a 98 % efficiency rating, which implies a heat transfer of 45 kWh (162 MJ).

The annual average water temperature in the storage tank, auxiliary tank, and the water mains are 40.1 °C (104 °F), 49.9 °C (122 °F), and 15.3 °C (60 °F), respectively. The solar thermal system raised the water temperature 24.7 °C of the 34.6 °C increase, or 71 %. Although this is not a precise calculation of the solar fraction, it is a reasonable ballpark estimate.[17]

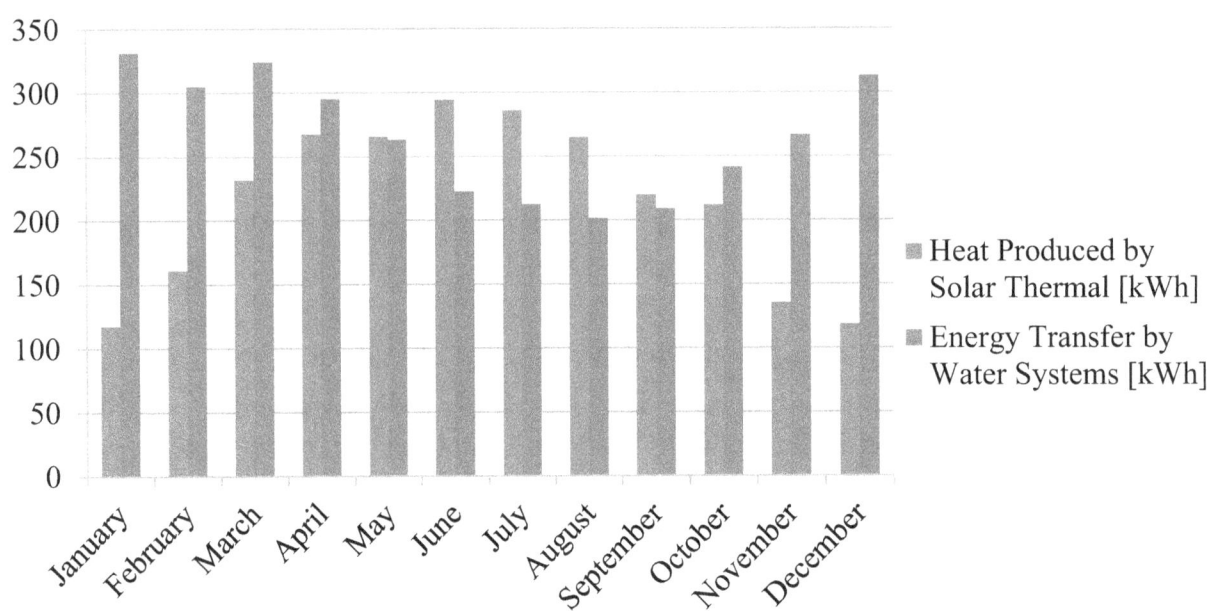

Figure 5-18 Heat Transfer to DHW Tank Water (kWh) - Monthly

The solar thermal system heats the water in a storage tank before the water enters the hot water tank. The temperature in the storage tank varies because of day-to-day weather variations. Figure 5-19 shows the water main temperature, temperature of the water exiting the storage tank, and water heater tank temperature. While the heat pump maintains the water temperature in the hot water tank, the storage tank temperature fluctuates due to the outdoor environment (temperature and solar radiance) and the water mains temperature. There are points in time at which the storage tank is significantly hotter than the hot water tank. In this situation, the water is tempered

---

[17] The average temperature in the auxiliary tank is 1 °C (1.8 °F) greater than the target temperature.

down to the target temperature of the hot water tank, trading some system efficiency for safety concerns.

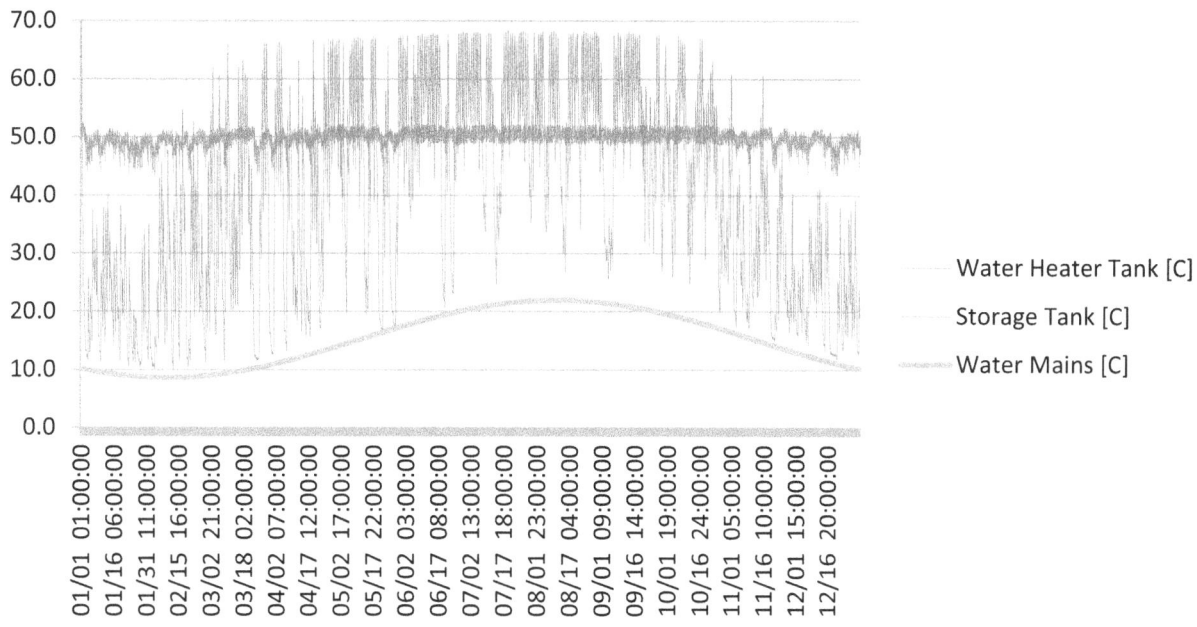

Figure 5-19 Water Main, Storage Tank, and Hot Water Tank Temperatures - °C

5.10   Room Temperature

Since the 1st floor is used as the control zone for the HVAC system, there may be a concern that the 2nd floor will have poor temperature control. As it turns out, the 2nd floor does vary slightly from the setpoint temperature. Figure 5-20 shows the average room temperature for the 1st floor and 2nd floor. The maximum difference in hourly average room temperature on the 2nd floor ranges between 1.8 °C (3.3 °F) lower and 1.7 °C (3.1 °F) greater than the 1st floor.

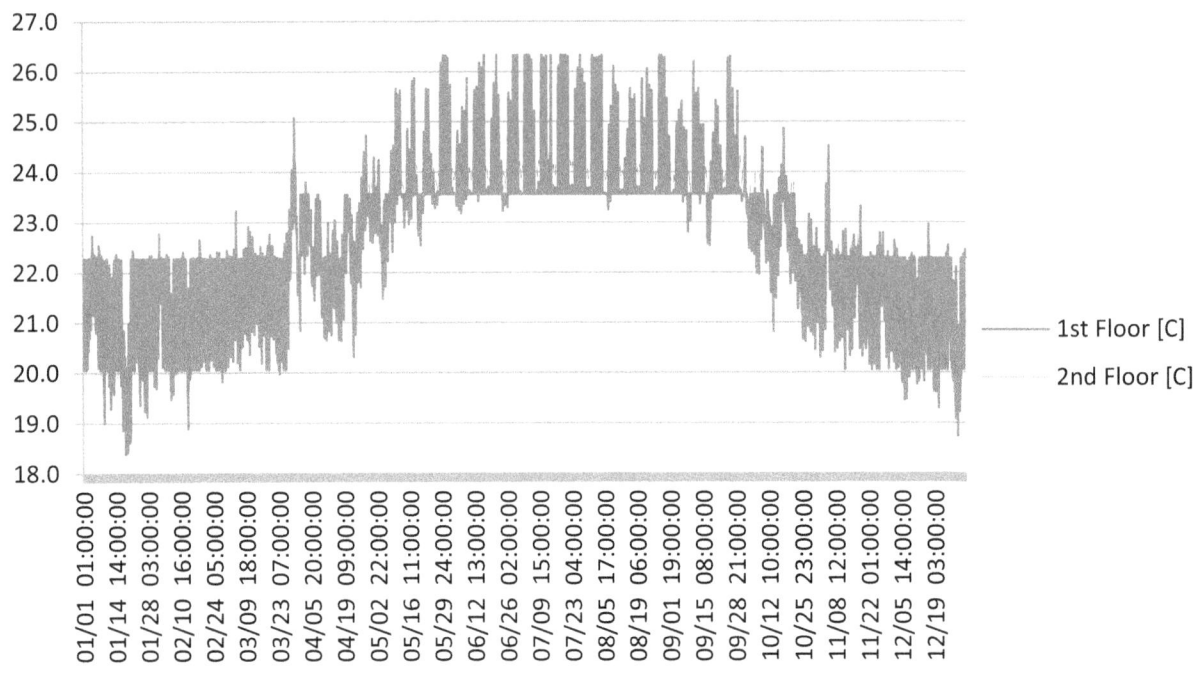

Figure 5-20 1st and 2nd Floor Mean Room Temperatures - °C

5.11 Humidity Levels

As can be seen in Figure 5-21, the humidity levels for each floor are restricted to always be below 60 % to ensure that the thermal comfort meets the *ASHRAE 90.2-2007 Standard* requirements. Both the 1st floor and the 2nd floor dehumidifiers are necessary to maintain the 60 % threshold, particularly during the cooling season. The humidity level is 0.4 percentage points lower for the second floor, on average. The 1st floor humidity level ranges from 6.3 percentage points lower to 2.6 percentage points higher. The basement realizes lower humidity levels during the summer than the occupied areas because there is no air infiltration into the basement.

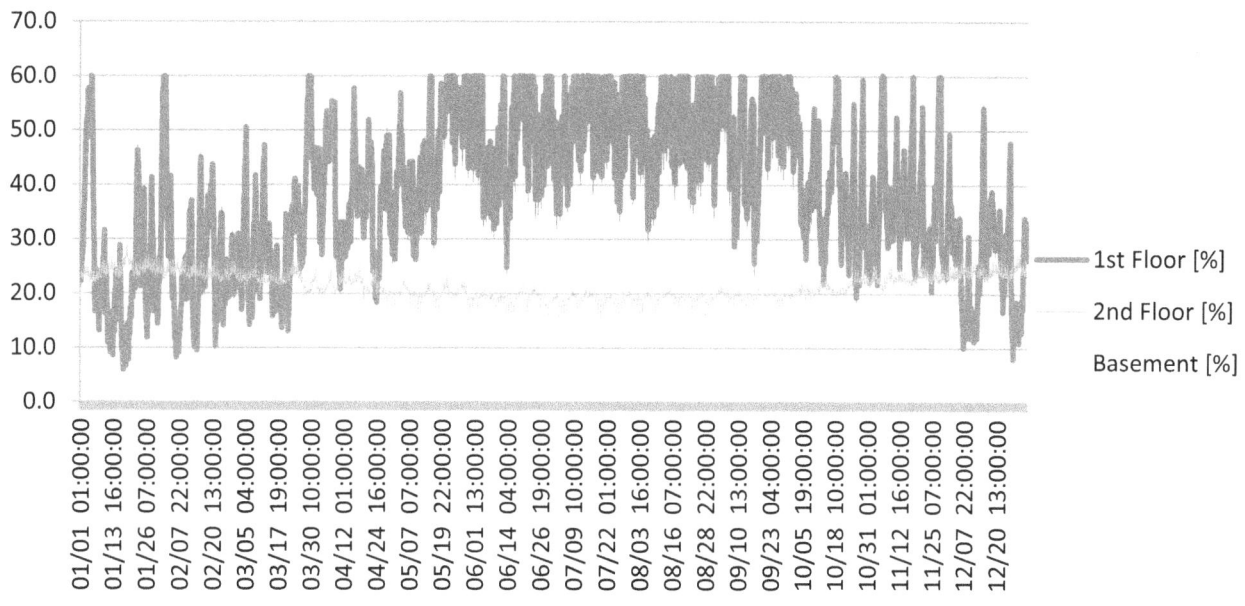

Figure 5-21 1ˢᵗ Floor, 2ⁿᵈ Floor, and Basement Relative Humidity Levels

Figure 5-22 shows that dehumidifiers are operated most frequently during the summer months to maintain the 60 % threshold. However, there are rare occasions during the winter months that dehumidification is necessary.

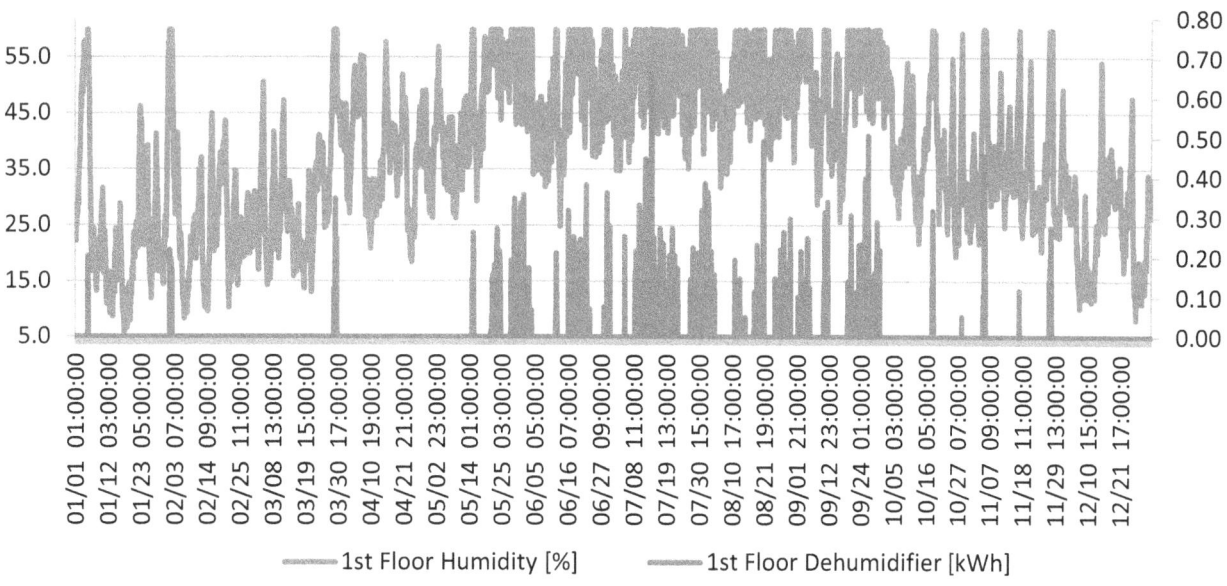

Figure 5-22 Hourly Humidity (%) and Dehumidifier Electricity Use (kWh) – 1ˢᵗ Floor

5.12 Thermal Comfort

The goal of the temperature and humidity control is to keep the thermal comfort levels in the building at acceptable levels. The 1st floor is rarely "not comfortable" based on *ASHRAE 55-2004* (160 hours) because it is the control zone. The 2nd floor is "not comfortable" over 4 times greater than the 1st floor (650 hours or 7.4 % of the year). Figure 5-23 shows that both floors have the greatest number of hours that are "not comfortable" during the colder months. The significant number of hours that the 2nd floor is "not comfortable" may be driven by the assumed setpoint temperatures, which may not closely match the conditions necessary to reach "comfortable" conditions.

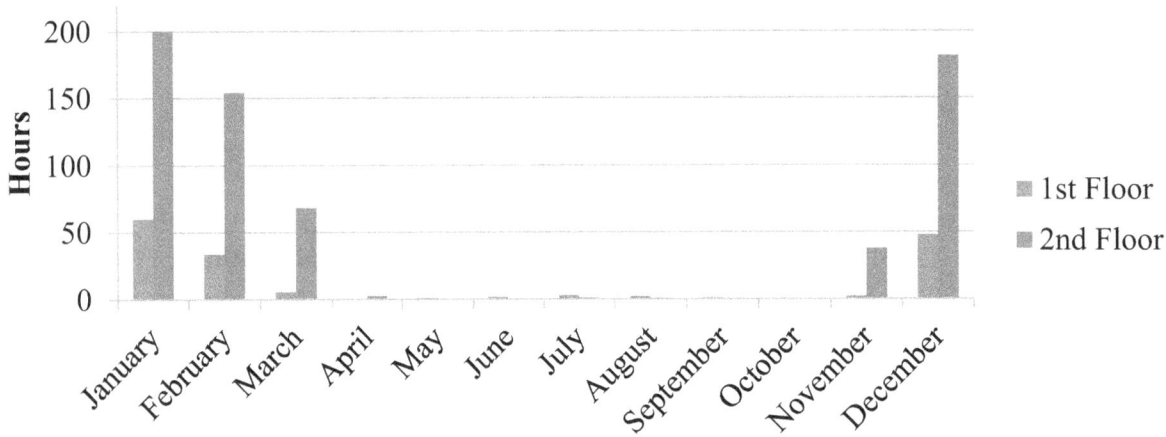

Figure 5-23 Simple *ASHRAE 55-2004* Not Comfortable - Hours

The "Time Set Point Not Met during Occupied Heating" on either floor is 406 hours while the "Time Set Point Not Met during Occupied Cooling" on either floor is 996 hours assuming a variation tolerance of 0.2 °C (0.4 °F). These values are greater than the 350 hour targets based on the design day conditions. The 1st floor performs at the target, with 168 hours not meeting the cooling setpoint and 270 hours not meeting the heating setpoint. The 2nd floor has 516 hours not meeting the heating setpoint, but has 945 hours not meeting the cooling setpoint.

Figure 5-24 shows that the amount of time the setpoint temperatures are not met in the control zone (1st floor) vary by month. The greatest amount of time that the heating setpoints are not met occur in January followed by December and February, which is logical given that those are the coldest months of the year. Similarly, the greatest amount of time that the cooling setpoint temperatures are not met occurs in July when the monthly cooling load is greatest.

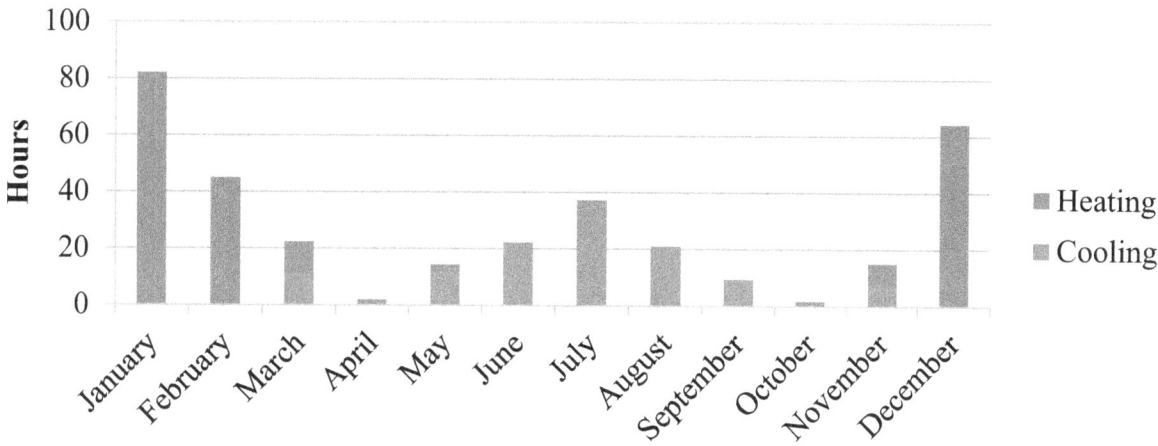

Figure 5-24 1ˢᵗ Floor Time Setpoint Not Met While Occupied - Hours

Figure 5-25 shows the amount of time the setpoint temperatures are not met on the 2ⁿᵈ floor varies by month, which is greater than those for the 1ˢᵗ floor because it is not the control zone. The greatest amount of time that the heating setpoint and cooling setpoint are not met occurs in September followed by August, July, and June. The HVAC system struggles most during the summer to maintain the 2ⁿᵈ floor temperature.

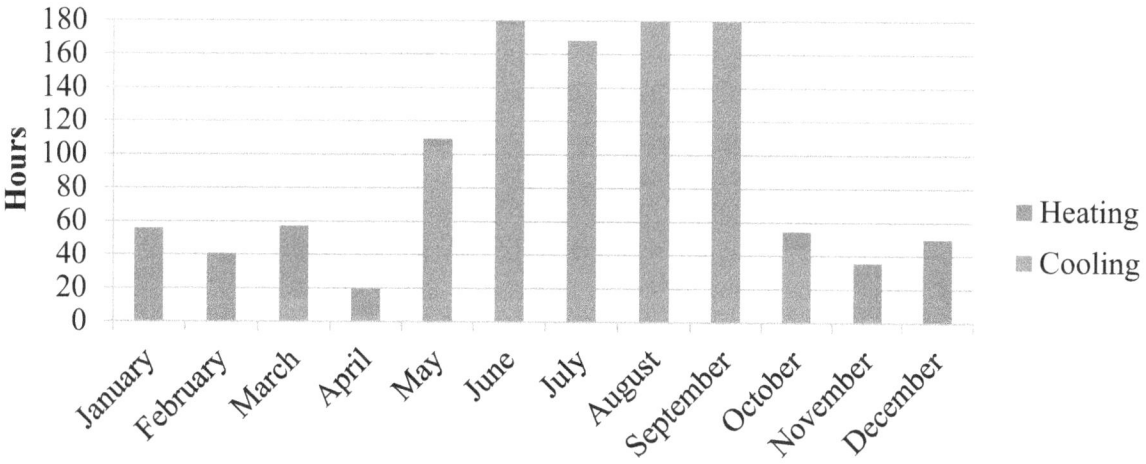

Figure 5-25 2ⁿᵈ Floor Time Setpoint Not Met While Occupied - Hours

5.13 Infiltration and Ventilation

Table 5-3 shows the average air flow rates (ACH) from infiltration and mechanical ventilation throughout the year. Mechanical ventilation has the greatest impact on outdoor air flow because the "tight" building envelope requires mechanical ventilation to maintain the appropriate outdoor air flow.

Table 5-3 Average Outdoor Air Flow during Occupied Hours - ACH

ACH	Ventilation	Infiltration	Total	Minimum Air Flow
1st Floor	0.125	0.054	0.179	0.090
2nd Floor	0.141	0.057	0.198	0.102

Figure 5-26 shows that the average mechanical ventilation is relatively constant throughout the year for both floors of the NZERTF. The 1st floor consistently has lower mechanical ventilation than the 2nd floor. For the 1st floor, 8 of the 12 months have an average ventilation rate of 0.13 ACH while the other months having slightly lower ventilation rates. For the 2nd floor, 11 of the 12 months have an average ventilation rate of 0.14 ACH while January has an average ventilation rate of 0.13 ACH.

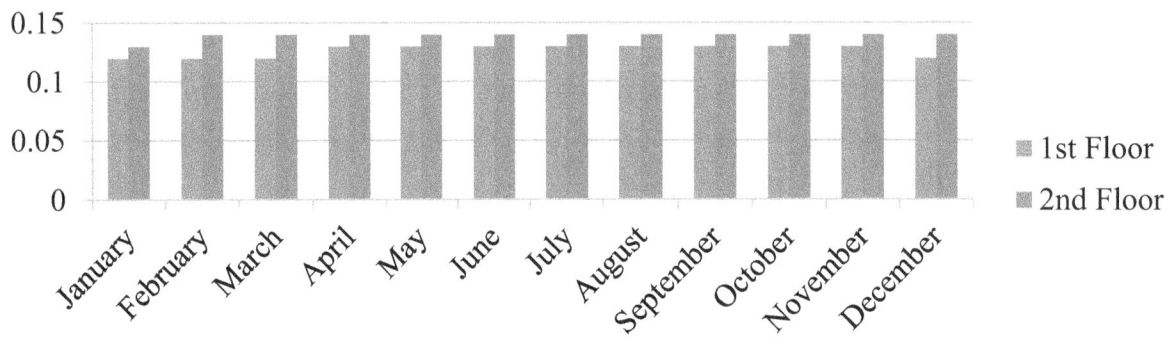

Figure 5-26 Mechanical Ventilation Rate by Floor - ACH

Figure 5-27 shows the very low average monthly infiltration rates by floor of the NZERTF. Either flow has a monthly average infiltration rate greater than 0.08 ACH. The 1st floor and 2nd floor have nearly identical average infiltration rates, with the 2nd floor having slightly more infiltration rates.

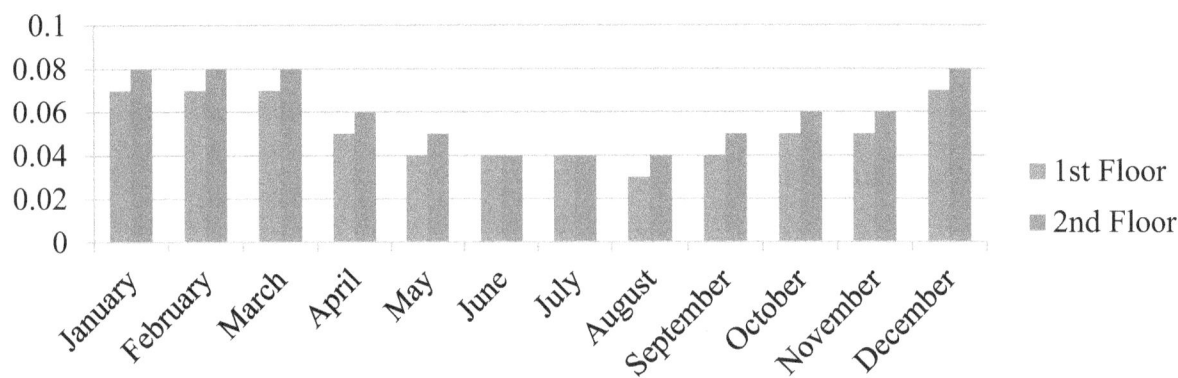

Figure 5-27 Infiltration Rate by Floor - ACH

Figure 5-28 shows the heat transfer, both gains and losses, resulting from infiltration on the 1st floor. The heat losses from infiltration are over 5 times greater than the heat gains from infiltration, which is likely driven by the greater heating degree days relative to cooling degree days. The heat transfer from infiltration for the 2nd floor has identical interpretations with slightly lower aggregate values as a result of less volume in the zone.

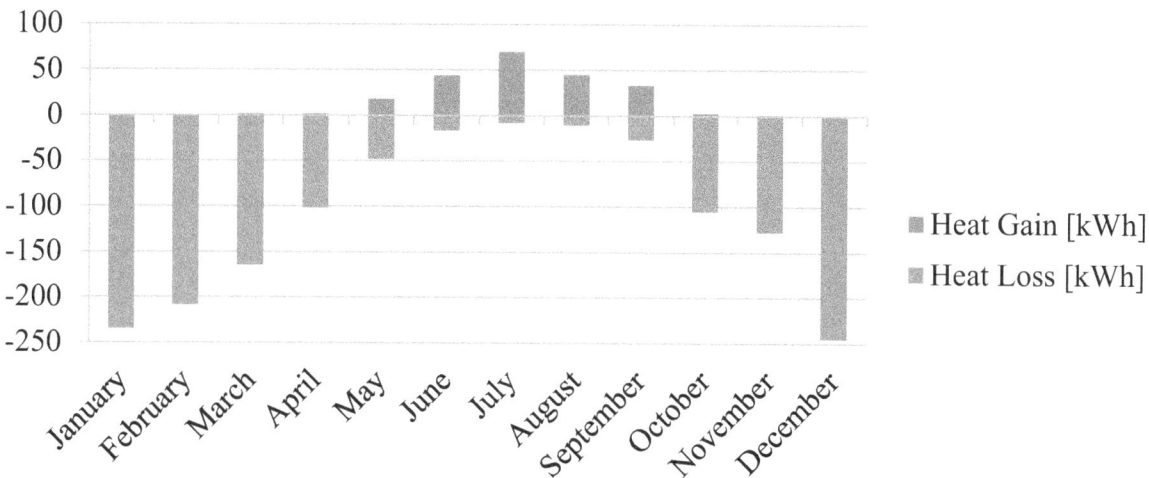

Figure 5-28 Infiltration Sensible Heat Transfer – 1st Floor (kWh)

Since the HRV system is the sole source of mechanical ventilation, its operation may work against the heating and cooling equipment by pulling in outdoor air that is warmer or colder than the target temperature of the control zone. Figure 5-29 shows that the mechanical ventilation increases the cooling load by a greater amount than the heating load, both in terms of monthly peak and on an annual basis. The more extreme the temperature outside, either low or high, the greater the impact of the HRV ventilation has on the thermal load of the NZERTF. This supports the hypothesis that the HRV works against the HVAC systems more often during the summer.

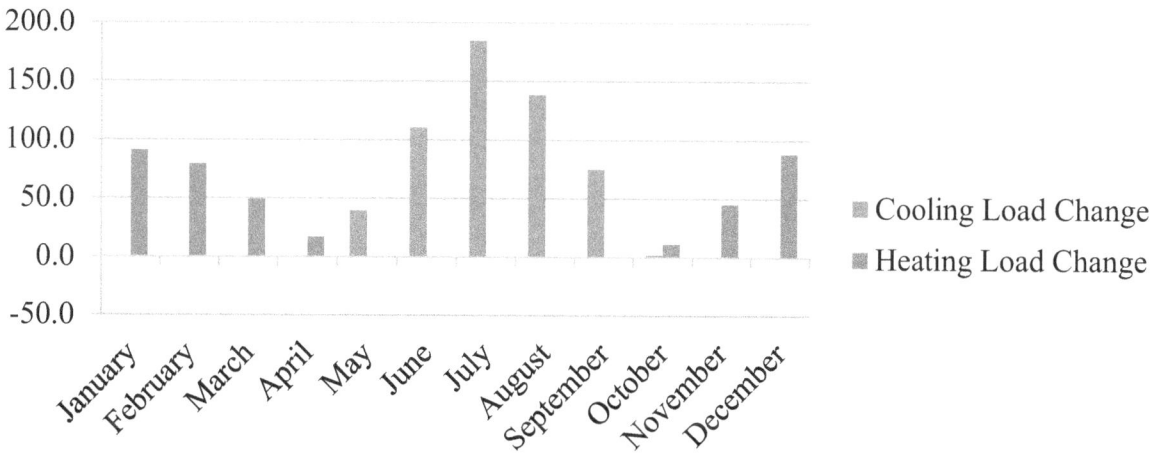

Figure 5-29 Load Changes from Mechanical Ventilation – 1st Floor (kWh)

61

6 Limitations

Whole building energy simulation software is limited in its abilities to estimate real world energy performance because it is difficult to control for all potential variables that can impact the thermal conditions in a building. There are a number of reasons that the estimated performance from the $E+$ model may vary from the actual performance of the NZERTF. I discuss a few below.

Discrepancies between the $E+$ model and the actual NZERTF design are the most prominent reasons for variations in energy performance. First, the $E+$ model does not include the stairway/foyer opening, which restricts thermal transfer between the floors and increases the conditioned floor area relative to the actual NZERTF design. The greater conditioned floor area will impact the thermal load needed to meet the thermostat set points as well as the minimum ventilation rates required for the HVAC equipment.

Second, air infiltration may not be accurately estimated in the $E+$ model. The infiltration rate due to building envelope leakage in the $E+$ model is simplified to an ELA based on the most recent air leakage test. However, this infiltration will not match actual infiltration the infiltration rate will differ depending on the outdoor weather conditions, relative pressure in the building, and location of the leakage areas. The $E+$ model currently does not control for these factors. The model does not account for any occupant-related infiltration, which will underestimate the total infiltration that will occur during the demonstration phase of the NZERTF project.

Third, building components may not perform at the manufacturer specifications. Solar PV production may vary drastically by weather and operating conditions. The simulation assumes optimal electricity production based on the TMY3 weather file. A single year may vary enough from the historical average to either increase or decrease solar PV performance. The solar PV system may be less productive than the specifications claim for some or all weather conditions, which would result in less electricity production than is estimated by the $E+$ model. There is minimal information available for the energy performance of the HVAC system, which could perform significantly better or worse than the simulation and drastically impact the NZERTF's energy performance. Some components of the HVAC and DHW systems are autosized in the $E+$ model even though in the actual NZERTF design the system will not be sized based on the simulation. Interior equipment and lighting may also differ from the expected performance.

Fourth, the monitoring and control of the NZERTF systems may change due to unforeseen circumstances. Monitoring and controlling every aspect of the NZERTF will be a complex and difficult task. Related problems could result in significant variations in electricity use depending on if systems are used more or less often than intended.

These limitations will be addressed whenever possible to improve the accuracy of the $E+$ simulation results. Updates based on the changes to the simulation and the related impacts on the results will be made as they occur.

7 Discussion and Future Research

The estimated performance of the NZERTF leads to several interesting results. First, if the house performs up to the energy simulations estimates, the NZERTF will perform well beyond its goal of net zero energy use even with an all-electric household. Second, renewable energy is the key to reaching net zero, both from electricity production by the solar PV system and pre-heating of DHW by the solar thermal system. Third, the plug loads become more important as a house becomes more efficient, which requires more efficient appliances as well as changes to occupant behavior to further reduce electricity use. Fourth, internal heat gains become more important in a "tight" building.

The current $E+$ model will be useful to assist in research in a number of areas, both directly related to the demonstration phase of the NZERTF project as well as broader research topics. The comparison of the $E+$ model performance relative to the actual performance of the NZERTF during the demonstration phase will be beneficial for two primary reasons. First, the model can be used to assist in fault detection for the building components in the NZERTF. Second, the actual performance can be compared to the $E+$ model to find areas of improvement for the $E+$ software.

The $E+$ model is best used to make relative comparisons of the NZERTF to alternative building designs. The most obvious comparison would be to an identical house that meets the Maryland state energy code for residential buildings. The energy performance savings can be compared to the additional costs of the efficiency improvements. Sensitivity analysis can be used to look at a number of interesting research areas. Sensitivity analysis can look at the incremental changes between the Maryland code compliant house and the NZERTF to determine which energy efficiency measures result in the greatest energy reductions, the most cost-effective measures, and how energy efficiency measures interact. The $E+$ model can also be altered to determine how additional energy efficiency measures would impact the energy performance of the NZERTF, including equipment that is already installed in the NZERTF. For example, each of the 3 geothermal heat pump systems could be included in the analysis to determine the impacts of systems on the NZERTF's energy performance. Occupant behavior can be altered to determine how sensitive the results are to the assumed occupant activity. Natural ventilation could be introduced by opening and closing windows based on the outdoor temperature to account for common human behavior.

References

AAON. "Modulating Hot Gas Reheat: The Split System Humidity Control Solution." www.aaon.com.

ANSI/ASHRAE Standard 55-2010. "Thermal Environmental Conditions for Human Occupancy." American Society of Heating, Refrigerating and Air-Conditioning Engineers, Inc. www.ashrae.org.

ASHRAE Fundamentals. 2009. American Society of Heating, Refrigerating and Air-Conditioning Engineers, Inc. www.ashrae.org.

ASHRAE Standard 90.2. 2007. American Society of Heating, Refrigerating and Air-Conditioning Engineers, Inc. www.ashrae.org.

ASHRAE Standard 62.2. 2010. American Society of Heating, Refrigerating and Air-Conditioning Engineers, Inc. www.ashrae.org.

Bosch Dishwasher. 2009. www.bosch-home.com.

Building Science Corporation. NIST NZERTF Zero Energy Performance. October 2009.

Building Science Corporation. 2009. 65 % Net Zero Energy Residential Test Facility Construction Documents Set.

Building Science Corporation. December 2011. NIST NZERTF Airtightness memo.

Christensen, D. and Winkler, J. December 2009. Laboratory Test Report for ThermaStor Ultra-Aire XT150H. NREL Technical Report NREL/TP-550-47215.

Department of Energy, Building Technologies Program, EnergyPlus energy simulation software Version 7.0.0, 2012, http://apps1.eere.energy.gov/buildings/energyplus/.

Energy Information Administration, Residential Energy Consumption Survey, http://www.eia.doe.gov/emeu/recs/.

Energy Information Administration, 2011, "Table 6. Class of Ownership, Number of Consumers, Sales, Revenue, and Average Retail Price by State and Utility: Residential Sector, 2010," http://www.eia.gov/electricity/data.cfm#sales.

Energy Star. Product databases for clothes washers, dishwashers, and refrigerators. http://www.energystar.gov/index.cfm?c=products.pr_find_es_products

Energy Star Home Sealing Specification. Version 1. 2001.

Engineering Toolbox. 2009. Material characteristics. http://www.engineeringtoolbox.com/

Environmental Protection Agency, 2012, Nissan Leaf EPA Fuel Economy Rating, http://www.fueleconomy.gov/feg/noframes/32154.shtml.

Everyday Green, 2012, Building Leakage Test of the Net Zero Energy Residential Facility.

Heliodyne Solar Hot Water Technical Specifications. www.heliodyne.com.

Heliodyne HPak Systems Installation Guide. www.heliodyne.com.

Hendron, R., Engebrecht, C., September 2010, Building America house simulation protocols, National Renewable Energy Laboratory, Building Technologies Program.

Hunter Douglas Duette FR Honeycomb Shades specifications. 2009. www.hunterdouglascontract.com/windowcoverings.

Lstiburek, Joseph. Building Science Corporation. BSI-030: Advanced Framing. Building Science Insights. http://www.buildingscience.com/documents/insights/bsi-030-advanced-framing/#F04

Omar, Farhad. Forthcoming. The NIST Net Zero Energy Residential Test Facility Occupancy Schedule and the Residents' Narrative. NIST Technical Note XXXX.

Pacific Northwest National Laboratory and Oak Ridge National Laboratory. December 2007. High Performance Home Technologies: Guide to Determining Climate Regions by County. Building America Best Practices Series.

Rheem HP50RH specifications. 2010. http://www.rheem.com/Products/tank_water_heaters/hybrid_electric.

Rudd, Armin. July 2010. Humidity Control Strategies. Residential Building Energy Efficiency Meeting 2010. Presentation.

Therrien Waddell. Appliance Letter of Transmittal. Wolf Product Specifications.

Serious Windows specifications. 2011. www.seriousmaterials.com.

Solar Photovoltaic System Letter of Transmittal. Therrien Waddell. 2012.

Sunpower 320 solar photovoltaic panel specifications. 2011. www.sunpower.com.

Sunpower 5000m inverter specifications. 2011. www.sunpower.com.

Ultra-Aire 65H Ventilating Dehumidifier specifications. www.Ultra-Aire.com.

Venmar AVS HRV EKO 1.5 specifications. 2009. http://www.venmar.ca/en/product/root-category/air-exchangers/venmar-avs/products/eko-15-hrv-_124.aspx?id_page_parent=775.

Winkler, J. and Christensen, D. August 2010. "Advanced Dehumidification Analysis on Building America Homes Using EnergyPlus." Xia Fang, NREL. Conference Paper. NREL/CP-550-48383.

Whirlpool Clothes Washer and Clothes Dryer specifications. 2009. www.whirlpool.com.

WILO Circulating Pump specifications. 2009. www.wilo.com.

www.ingramcontent.com/pod-product-compliance
Lightning Source LLC
Chambersburg PA
CBHW081835170526
45167CB00007B/2812